Sustainable Management of Cordyceps

This book examines the challenges of sustainably managing and conserving *Cordyceps sinensis*, a rare species of fungus largely grown in Tibet, currently on the brink of extinction.

As one of the most expensive commodities in the world, particularly valued for its medicinal properties in China, the price of Cordyceps has risen by over 900% since the 1970s. This has made it a very lucrative resource for farmers, many of whom are struggling to produce sufficient food to sustain themselves. Naturally, this has led to overharvesting and, coupled with the impacts of climate change, the crop itself is now at risk. Rarely discussed in Western literature, this book provides a novel examination of Cordyceps, looking into the necessary changes needed to sustainably manage and conserve this important crop. Drawing on extensive field work conducted in Qinghai-Tibet, the book analyzes the supply chain, identifying key issues around production and considering the role and impact of relevant stakeholders. It discusses the necessary changes needed for a sustainable supply chain, particularly to stop long-term overharvesting. The book then discusses the role of policy and the institutional management of this resource in China, as one of the main producers and consumers. It analyzes current policy instruments and argues for a more coherent policy which is better orientated towards conservation and sustainable management, rather than solely market regulation.

This book will be of great interest to students and scholars of natural resource management, environmental conservation, environmental policy, and sustainable supply chain management.

Jiping Sheng is a Professor in Food Economics and Management and Food Science at Renmin University of China. She has been working in these fields for nearly 30 years with more than two hundred papers and twenty books published.

Ksenia Gerasimova is a Professor in Public Policy at the Higher School of Economics, Moscow, Russia, and is CEENRG Fellow in Land Economy at the University of Cambridge, UK. She is the author of *NGO Discourses in the Debates on Genetically Modified Crops* (Routledge, 2017).

Earthscan Studies in Natural Resource Management

Resource Communities
Past Legacies and Future Pathways
Kristof Van Assche, Monica Gruezmacher, Lochner Marais, and Xaquin Perez-Sindin

English Urban Commons
The Past, Present and Future of Green Spaces
Christopher Rodgers, Rachel Hammersley, Alessandro Zambelli, Emma Cheatle, John Wedgwood Clarke, Sarah Collins, Olivia Dee and Siobhan O'Neill

Governing Natural Resources for Sustainable Peace in Africa
Environmental Justice and Conflict Resolution
Edited by Obasesam Okoi and Victoria R Nalule

Capacity-Building and the Water-Energy-Food Nexus
Rethinking Integration in the Asia-Pacific
Maureen Papas

Governing the Palm Oil Industry
Perspectives from Southeast Asia and Latin America
Edited by Patrick O'Reilly and Helena Varkkey

Researching Institutions in Natural Resource Governance
Methods and Frameworks
Edited by Fiona Nunan

Sustainable Management of Cordyceps
Supply Chains and Resource Management Policies
Edited by Jiping Sheng and Ksenia Gerasimova

For more information about this series, please visit: www.routledge.com/books/series/ECNRM/

Sustainable Management of Cordyceps

Supply Chains and Resource Management Policies

Edited by Jiping Sheng and Ksenia Gerasimova

First published 2025
by Routledge
4 Park Square, Milton Park, Abingdon, Oxon OX14 4RN

and by Routledge
605 Third Avenue, New York, NY 10158

Routledge is an imprint of the Taylor & Francis Group, an informa business

British Library Cataloguing-in-Publication Data
A catalogue record for this book is available from the British Library

ISBN: 978-1-032-54815-9 (hbk)
ISBN: 978-1-032-55252-1 (pbk)
ISBN: 978-1-003-42975-3 (ebk)

DOI: 10.4324/9781003429753

Typeset in Times New Roman
by Apex CoVantage, LLC

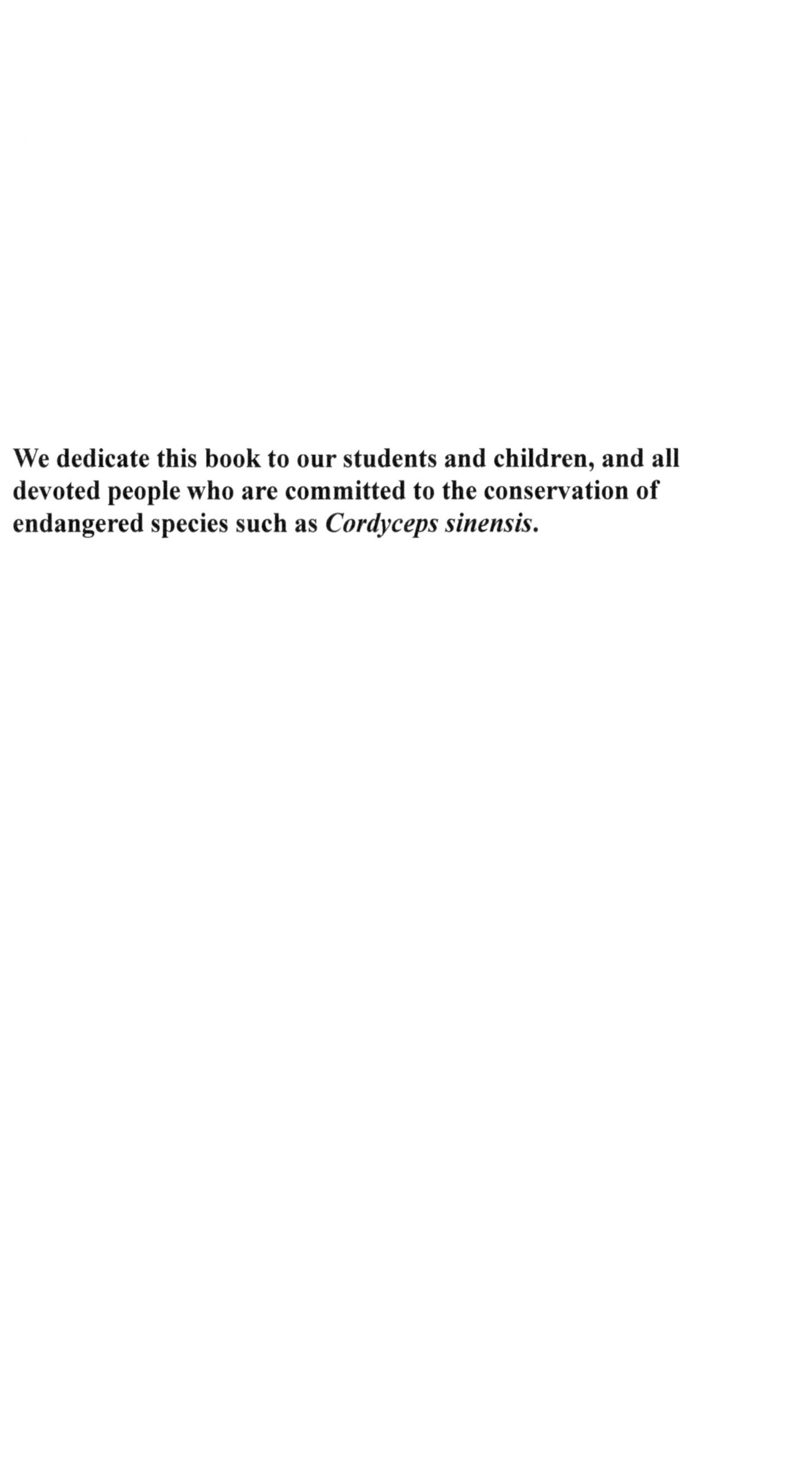

We dedicate this book to our students and children, and all devoted people who are committed to the conservation of endangered species such as *Cordyceps sinensis*.

Contents

Figures

Tables

Contributors

This book is based on the reports formed from our multiple surveys and the materials from graduate students' theses. The research work has undergone years of effort with many people involved. The authors listed here mainly refer to those who participated in the organization of the content in the eleven chapters of this book; some researchers who participated in the surveys have not been listed here.

Prof. Jiping Sheng was fortunate to receive a grant project from the Ford Foundation in 2013, 'Research on Sustainable Development of Agricultural By-products Based on Environmental Protection and Ensuring Income of Farmers and Herdsmen—A Case Study of the *Cordyceps sinensis* Production Area in the Qinghai-Tibet Plateau' (0130–1447). She organized a research team composed of teachers and graduate students from Renmin University of China, China Agricultural University, and the High School Affiliated to Renmin University of China, to conduct surveys and investigations. By the time the project was completed in 2017, multiple research reports were generated, and the first book in China on the management of *Cordyceps sinensis*, titled *Resource Management and Sustainable Development of Cordyceps sinensis*, was published by China Agricultural University Press in the same year. Even though the project has ended, this work has continued, and during the COVID-19 pandemic, the consumption of *Cordyceps sinensis* was in focus. In 2023, the research group went to Gansu for surveys. Doctoral student Jinshuo Zhang also organized the research conducted in Nyingchi, Tibet in 2023 and included it in this book. Professor Ksenia Gerasimova participated in the surveys in Qinghai and TAR, and also conducted additional research on *Cordyceps sinensis* in Bhutan, Nepal and Russia.

The authors and the name of their institution are as follows:

Jiping Sheng, Professor in the School of Agricultural Economics and Rural Development, Renmin University of China.

Ksenia Gerasimova, Professor in Public Policy, Higher School of Economics, Moscow; CEENRG Fellow in Land Economy, University of Cambridge. She served as Invited Lecturer in the School of Agricultural Economics and Rural Development, Renmin University of China, Beijing in 2021. She joined Shenzhen MSU-BIT University in March 2024.

Lin Shen, Professor in the College of Food Science and Nutritional Engineering, China Agricultural University.

Jiang Zhao, Doctoral student in the School of Agricultural Economics and Rural Development, Renmin University of China.

Shenghang Wang, Doctoral student in the Law School, Renmin University of China (Assistant Professor in School of Law, Shandong University).

Jinshuo Zhang, Doctoral student in the School of Applied Economics, Renmin University of China.

Jiahao Shi, Doctoral student in the School of Geography and Ocean Science, Nanjing University, China.

Yiyuan Miao, Doctoral student in the School of Agricultural Economics and Rural Development, Renmin University of China.

Wenfan Su, Doctoral student in the School of Agricultural Economics and Rural Development, Renmin University of China.

Huiqi Song, Doctoral student in the School of Agricultural Economics and Rural Development, Renmin University of China.

Mengyao Diao, Master's student in the School of Agricultural Economics and Rural Development, Renmin University of China.

Wenjie Long, Master's degree student in the School of Agricultural Economics and Rural Development, Renmin University of China.

Yiyun Wen, Master's degree student in the School of Agricultural Economics and Rural Development, Renmin University of China.

Xinyi Xu, Master's degree student in the School of Agricultural Economics and Rural Development, Renmin University of China.

Yijing Xin, Master's degree student in the School of Agricultural Economics and Rural Development, Renmin University of China.

Shengye Shen, Undergraduate student at Smith College, United States.

Shengming Shen, Undergraduate student at the School of Life Sciences, Tsinghua University, Beijing, China.

Ekaterina Zueva, Master's degree student in Geography, Shenzhen MSU-BIT University.

Foreword

Cordyceps sinensis, an esteemed medicinal fungus unique to the Qinghai-Tibet Plateau, is renowned alongside ginseng and deer antler as one of the 'Three Treasures of Chinese Medicine.' With over 98% of the global cordyceps supply originating from China, the industry has become a cornerstone of the economies in the provinces of Qinghai and Tibet. The nutritional and medicinal value of cordyceps has been widely recognized due to extensive research conducted both domestically and internationally, significantly bolstering the industry's growth and catalyzing the development of associated regional industries, thereby establishing a model where regional economies are bolstered by local proprietary industries.

Three key attributes stand out in the cordyceps industry: Firstly, substantial scientific and technological breakthroughs have been achieved, yet substantial room for advancement remains. China's advantageous geographical and climatic conditions have facilitated natural science research on cordyceps since the 1970s, with ecological studies ranking among the global forefront. However, there are still gaps in foundational theoretical research areas such as pharmacology and artificial cultivation that necessitate further exploration.

Secondly, the industry has experienced rapid development, positively impacting the local populace. Since the 21st century, the market price of cordyceps has seen a dramatic increase, from several thousand to tens of thousands, markedly elevating the standard of living for farmers and herders in producing regions. Concurrently, a diverse array of products, including cordyceps-containing lozenges, capsules, oral suspensions, and health wines, has emerged in the marketplace, propelling the cordyceps industry to a valuation approaching one trillion.

Thirdly, the lack of research in the industrial economics domain has impeded the industry's progress. Despite China's leading role in natural science research on cordyceps, the study of the industry's economic dynamics remains inadequate, with most research being macro-level analyses rather than focusing on meso-level supply chain dynamics and micro-level industry-specific inquiries. This gap has been compounded by the escalating economic value of cordyceps, which has led to overexploitation, causing significant ecological degradation and a rapid depletion of cordyceps resources, thereby hindering the industry's further advancement.

During my tenure at the Ministry of Agriculture, I was instrumental in organizing research initiatives on the cordyceps industry, and Professor Jiping Sheng from

China Agricultural University (now in Renmin University of China), an eminent scholar, contributed extensively to these efforts. Professor Ji Ping's research team, specializing in cordyceps, has conducted additional profound research into China's cordyceps industry, examining the supply chain and the perspectives of industry stakeholders, offering a novel perspective that enriches and advances the field of cordyceps industry research.

This book compiles the meticulous research efforts of Professor Sheng's team over nearly 10 years, encapsulating the historical context, geographical distribution, contemporary scientific investigation, and culinary utilization, food safety policy and management rules of cordyceps. It provides comprehensive insights into the cordyceps industry, its various operational stages, the personnel involved, and the legal and regulatory framework that governs it. The publication of this treatise further enriches the scholarly discourse on the cordyceps industry and is a treasure trove of knowledge for any aficionado of this remarkable medicinal fungus.

Lijian Zhang,
PhD, Researcher, Vice President of the
Chinese Agricultural Academy of Sciences

Foreword

I am honored to write this foreword for the groundbreaking collaborative work of Professor Jiping Sheng and Professor Ksenia Gerasimova. This book marks a significant milestone as the first to explore the management and policy aspects of *Cordyceps sinensis* in English.

Professor Jiping Sheng is a distinguished academic whose deep expertise in food safety management has earned her a reputation as a leader in the field of food science and management. Her unique insights and multidisciplinary approach have not only advanced the field but have also inspired students and colleagues to pursue academic excellence. As an academic leader in this field, her research has contributed substantially to our understanding of food systems and their sustainability.

The collaboration between Professor Jiping Sheng and Professor Ksenia, a renowned scholar from the University of Cambridge, is a testament to our school's commitment to fostering partnerships with the world's top expert. Professor Ksenia's participation in some cooperative studies with our school speaks to her expertise and the value she brings to these endeavors.

Cordyceps sinensis, also known as Chinese caterpillar fungus, has long been cherished for its remarkable health benefits and medicinal properties. The increasing demand for this rare and valuable resource has raised concerns about its sustainability, making the topic of resource management a critical issue of our time. This book provides a comprehensive analysis of the current situation of *Cordyceps sinensis*, examining the ecological, economic, and social dimensions of its harvest and utilization.

The authors have meticulously researched the complex interplay of factors that affect the sustainability of *Cordyceps sinensis*, drawing on data and insights from various stakeholders, including local communities, researchers, and policymakers. The book explores the potential impacts of different management strategies on ecosystems and the role of market mechanisms, community-based initiatives, and regulatory frameworks in promoting sustainable resource management.

The recommendations provided in this book aim to ensure the long-term viability of *Cordyceps sinensis* while preserving the ecological integrity of its habitats, contributing to the well-being of present and future generations. It is a significant work that will undoubtedly inspire and guide stakeholders in their efforts to achieve sustainable resource management of this valuable resource.

As we anticipate the publication of this important book, we celebrate the achievements of Professor Sheng and Professor Gerasimova. Their collaborative efforts have produced a scholarly work that addresses a critical issue in the fields of food safety management and resource conservation. This book is a testament to their expertise, commitment, and dedication to advancing knowledge and practice in their respective fields.

I highly recommend this book to policymakers, researchers, and practitioners interested in food safety management and resource conservation. It is a valuable resource that offers practical solutions and policy recommendations for achieving sustainable resource management of *Cordyceps sinensis* worldwide.

Huanguang Qiu,
PhD, Professor, Dean of the School of Agricultural Economics
and Rural Development, Renmin University of China.

Acknowledgments

We appreciate the financial support received from the Ford Foundation to complete a research project titled ‘Research on Sustainable Development of Agricultural By-products Based on Environmental Protection and Ensuring Income of Farmers and Herdsmen—A Case Study of the *Cordyceps sinensis* Production Area in the Qinghai-Tibet Plateau’ (0130–1447).

Professor Ksenia Gerasimova was a receiver of Top-Foreign Expert, China (in Sustainable Agriculture) Award in 2021–2022.

We thank the School of Advanced Studies, a ‘greenfield research body’ of Tyumen University, and its director Professor Andrey Scherbenok for assistance with the research conducted in Siberia, Russia.

We thank our research team, including our graduate students, for joining us as we hiked through mountains and rivers to conduct research in impoverished areas, interviewed farmers and herders, and gave lectures in Yi Ethnic minority mountain schools! The students returned to Renmin University of China to raise funds for clothing donations to the children in the mountainous regions. We are also grateful to Professor Yala Peng and her husband, Professor Jianfeng Peng, for their generous donation of a large number of computers and desks and chairs to Liangshan Yi Ethnic Primary School.

We thank the local leaders, farmers and herders for accompanying us on our research and providing us with information and data. We thank the experts and peers who provided valuable suggestions for our project!

We express our gratitude to the following individuals and more: Yang Wang, Director of the Sichuan Meigu Dafengding National Nature Reserve Administration Bureau; Xiaobao Yao, Director of the Qinghai Cordyceps Association; Xiaojun He, General Manager of Mount Everest Pharmaceuticals; Chengxiang Yang, Director of Mount Everest Pharmaceuticals; Yuhua Wang, Mount Everest Pharmaceuticals; Yuling Li, Professor at the College of Animal Husbandry and Veterinary Science, Qinghai University; Lixin Wei, Director of the Northwest Institute of Plateau Biology, Chinese Academy of Sciences; Yueyuan Wang, Qinghai Zhou Long Biotechnology Co., Ltd; Jie Meng, an expert in intercultural communications, Beijing; Binguo Yu, an expert and a friend in Beijing.

Introduction

Jiping Sheng and Ksenia Gerasimova

Cordyceps sinensis, commonly known as Winter Worm Summer Grass or Chinese caterpillar fungus, is a rare and highly valued medicinal fungus that is native to China. This unique fungus has been used in traditional Chinese medicine (TCM) for centuries, renowned for its supposed healing properties and as a tonic for various ailments. It is believed to have benefits for the immune system, respiratory health, kidney function, and as an overall energy booster. In TCM, *Cordyceps sinensis* is considered a yang-invigorating herb that can raise body temperature and increase vitality.

Harvesting *Cordyceps sinensis* is a challenging and labor-intensive process, as it requires finding the infected caterpillars deep in the mountains, often at high-altitude areas, mainly distributed in the Qinghai-Tibet Plateau region. Due to its rarity and high demand, *Cordyceps sinensis* can be extremely expensive, making it a lucrative industry in regions where it is naturally occurring.

In recent years, the annual output value of *Cordyceps sinensis* products has been continuously increasing, making it one of the most important industries in the production areas and a major source of income for local poor farmers and herdsmen. After 2000, the market demand for *Cordyceps sinensis* soared, causing its price to rise by tens of times in just a few years, reaching more than 200,000 yuan per kilogram in 2013. By 2023, mature activities such as the Fresh Grass Festival and artificial cultivation of *Cordyceps sinensis* had developed, further expanding the market. Annual sales volumes reached 100–200 tons, with 80% sold in China (including Hong Kong, Macau, and Taiwan) and exported to countries such as Europe, America, Japan, and Southeast Asia. During the annual collection period, most farmers and herdsmen in the *Cordyceps sinensis*-producing areas join the collection team, and many farmers from surrounding areas are also attracted, which has had a significant impact on the sustainable development of the industry. How to achieve sustainable development and utilization of *Cordyceps sinensis* has become an urgent issue that needs to be addressed.

Against this background, our team took on a research project funded by the Ford Foundation in 2013, 'Research on Sustainable Development of Agricultural By-products Based on Environmental Protection and Ensuring Income of Farmers and Herdsmen—A Case Study of the *Cordyceps sinensis* Production Area in the Qinghai-Tibet Plateau' (0130–1447). The goal was to understand the industry

DOI: 10.4324/9781003429753-1

situation of *Cordyceps sinensis* through research, taking into account both environmental protection and farmers' and herdsmen's income, and seeking a path to sustainable development. We established a research team and conducted eight large-scale investigations from 2013 to 2017 during the five years of project implementation. We traveled tens of thousands of kilometers, visited Sichuan Liangshan Yi Autonomous Prefecture, Gansu, Qinghai, Tibet, Yunnan, as well as large and medium-sized cities such as Beijing, Shanghai, and Qingdao for consumer research, and visited *Cordyceps sinensis* consumers such as the China Polar Research Center. We investigated the resource status, its natural conditions, the overall ecological environment, historical background and local lifestyle, and communicated with local farmers and government officials. Through technical training for farmers and herdsmen, we helped to curb the current situation of exploitative collection of *Cordyceps sinensis* resources and maintain the sustainable use of *Cordyceps sinensis* resources. At the same time, we spoke with farmers and herdsmen about their interest and skills in environmental protection and encouraged them to learn about the production and development of ecological and organic foods, helping them find a sustainable alternative to *Cordyceps sinensis* collection. We also studied consumer perceptions of *Cordyceps sinensis* from the consumption end, especially the series of changes that occurred throughout the *Cordyceps sinensis* industry chain from the time it was removed from the list of medicinal and food-sourced substances by the Chinese Health Commission in 2015.

After the project ended, the work continued. In 2022–2023, we conducted two surveys in Qinghai and Tibet, visited the world's largest *Cordyceps sinensis* trading market, the Jiuying *Cordyceps sinensis* Trading Market, and the New Millennium *Cordyceps sinensis* World, communicated with sales agents, and learned about the current supply and demand situation in the *Cordyceps sinensis* market. Since the COVID-19 pandemic, due to *Cordyceps sinensis*' efficacy in boosting immunity and other benefits, there has been a significant increase in demand, with prices for different sizes of *Cordyceps sinensis* in 2022 increasing by 15% to 35% compared to the previous year. In April 2023, we also visited Qinghai Zhufeng Pharmaceutical Company and the Qinghai *Cordyceps sinensis* Association to learn about their work progress in the collection, sales, and safety management of *Cordyceps sinensis*, such as changes in market regulatory measures like permits for digging. On joining the team in 2017, Professor Ksenia Gerasimova contributed her findings from her fieldwork research in Bhutan in 2015 and in Russia in 2023, where she obtained a wealth of first-hand data. She has also taken part in field visits to Qinghai and TAR since then.

In recent years, overharvesting and the destruction of natural habitats have led to a decline in wild populations, prompting efforts to cultivate the fungus commercially. The cultivation process involves carefully inoculating host insects with the *Cordyceps sinensis* spores under controlled conditions to produce the valuable fungus in a sustainable manner. Some *Cordyceps sinensis*-producing areas such as Qinghai Province have adhered to the concepts of ecological priority and green development, focusing on the two main themes of 'increasing greenness' and 'increasing income,' continuously promoting high-quality development of the

Cordyceps sinensis industry, and striving to cultivate a complete industrial chain from collection to processing and sales, continuously increasing the income of farmers and herdsmen.

Cordyceps is a global resource that we should cherish and utilize sustainably and scientifically. While harvesting wild Cordyceps, collectors should pay attention to protecting the ecological environment, and artificial cultivation of the *Cordyceps sinensis* can ensure protection of the wild pool of this resource.

Part I

Introduction to Cordyceps

A true treasure of the Qinghai-Tibetan mountainous area, Cordyceps have been known as a powerful traditional Chinese medicinal treatment. In the 21st century, Cordyceps caught the attention of wider international audiences, and its long history hopefully ensures all possible efforts are being made to explore its unique medicinal qualities and to protect its habitat, and to support local people, who rely on harvesting wild Cordyceps.

While participating at the Leaders' Summit on Climate in 2021, President Xi Jinping spoke about the Chinese framework of sustainability, which is an amalgamation of the traditional values about the Nature that inspired the conceptualization of the Chinese equivalent of Sustainable Development—the Ecological Civilization, and the modern commitment to participate in multilateral efforts to address climate change (Xinhua, 2021).[1] Like in other parts of the world, the issue of balancing socio-economic development with environmental conservation is a wicked problem, as already recognized by scientists, the public, and policy-makers, including the honorable Chinese leader. Sustainable management of Cordyceps has many layers of wickedness—it is difficult to simultaneously protect this endangered species when it provides the main alternative to traditional rural activities, but the people are already suffering from climate change.

In theory, scientists can conceptualize how to save important species, but in practice there are also gaps in the information and not a one-size-fits-all strategy, such as creating a nature reserve, which is not enough, as several factors contribute to rare species' habitat deterioration, including climate change and destructive influences from human development, particularly agriculture (Sutherland, 2022).[2] We hope that this monograph will summarize the existing knowledge and push additional international efforts to protect and cherish Cordyceps.

Notes

1 Xinhua. (2021). Remarks by Chinese President Xi Jinping at Leaders' Summit on Climate, 4 April 2021. www.xinhuanet.com/english/2021-04/22/c_139899289_2.htm

2 Sutherland, W.J. (2022). Transforming conservation: A practical guide to evidence and decision making. Open Book Publishers: Cambridge. https://doi.org/10.11647/OBP.0321

DOI: 10.4324/9781003429753-2

1 What Is Cordyceps?

Jiping Sheng, Mengyao Diao, Yiyuan Miao, and Ksenia Gerasimova

Introduction

In this chapter, we explain the morphology of Cordyceps and the particularities of its habitat. This will serve as a foundation for discussing further policy interventions in conservation strategies and market regulation.

To remind, an ecological system is understood as a closed system of ecological community and its controlling natural environment based on nutrient cycles (Borman & Likens, 1970). Identifying the key variables in its functioning is an important method to assess prosperity and resilience of the total ecosystem and individual segments. The concept of ecosystem resilience has been described as the internal persistence of the system to absorb changes. But management of ecosystems under disturbance can have two distinct approaches. One is based on stability of the system, which describes how fast and at what price the ecosystem can return to an equilibrium state after the disturbance. The second approach is about resilience that is more flexible, open and can emphasize heterogeneity and increase the capacity of the system to accommodate future disturbances (Holling, 1973, pp. 17, 20). In practice, survival or distinction of a species is defined by a combination of random events and predetermined conditions. The advice of Holling (1973) to keep all the options open in ecosystem management and work with changing human ignorance is still valid. But better understanding of the linkage and relationships within the ecosystem can be beneficial to increase its capacity to withstand a specific disturbance event.

As with other fungi, Cordyceps are heterotrophic organisms that depend on other organisms for supplementing nutrients through the microbiological breakdown of organic materials and their absorption. Through such a mechanism, fungi are highly integrated in the ecosystem, as they require interrelations with other organisms. This interrelation is called symbiosis. In general, fungi form disjunctive and conjunctive, antagonistic, neutral and mutual symbiosis (Cooke, 1979, p. 1). For Cordyceps, it is an antagonistic symbiosis, as a bat moth, the host, is invaded and destroyed as a result. Cordyceps cannot develop without a host insect, and this means that they are highly sensitive to any change in their environmental complex. From such perspective, the assessment of the 'well-being' of Cordyceps can be used as a litmus test to measure the threshold of the whole

DOI: 10.4324/9781003429753-3

ecosystem and its ecological robustness. Understanding the key factors changing the balance in the complex can help first to identify and then to alleviate the most damaging anthropogenic interferences and ensure the implementation of efficient conservation efforts.

Cordyceps is a general term for fungi that include the well-known *Cordyceps sinensis*. Species of Cordyceps are named after morphological characteristics, host, and geographic location. So far, more than 400 species of Cordyceps fungi have been identified in the world, and *Cordyceps sinensis* refers to a species which is mainly found in the Qinghai-Tibetan Plateau (Yue & Ye, 2013). *Cordyceps militaris* can be found outside the Himalayas, in temperate forests of North America (Canada, the Appalachian Mountains and forested parts of Mexico) and Europe (WMS, 2022).

Cordyceps sinensis, also known as Chinese Cordyceps, is called *Dong chong xia cao* in Chinese (it means 'a winter worm turns a grass in summer') and *yartsa gunbu* in Tibetan (Figure 1.1a and 1.1b). It was first recorded in the ancient encyclopedia of traditional Chinese medicine, *Yue Wang Yao Zhen* (《月王药诊》), in the 8th century AD (Buddhist Digital Resource Centre, 2006). For its high nutritional value and good medical efficacy, it is also known as 'the Best Grass in the World' and listed as one of China's three major tonics together with ginseng and Cornua Cervi Pantotrichum. Its medicinal value and nourishing effects rank first among the three famous tonics, with the effect of strengthening kidneys and lungs, stopping bleeding, and removing phlegm. It is used in traditional medicine to treat a chronic cough and hemoptysis, impotence, waist and knee pain and many other diseases (ibid.).

Cordyceps is a parasite that invades larvae of the ghost bat moth (Hepalius humuli), which grows throughout its host's body in summer, thus it is called 'winter worm and summer grass' in Chinese. Every March, a bat moth comes to the grassy slope near the snow line of the alpine meadow shrubs, at an altitude of 3000–5000 meters, and lays the eggs in the clods. Hatched bat moth larvae crouch in the damp and warm soil over the winter and feed on the tender roots and succulence of wild

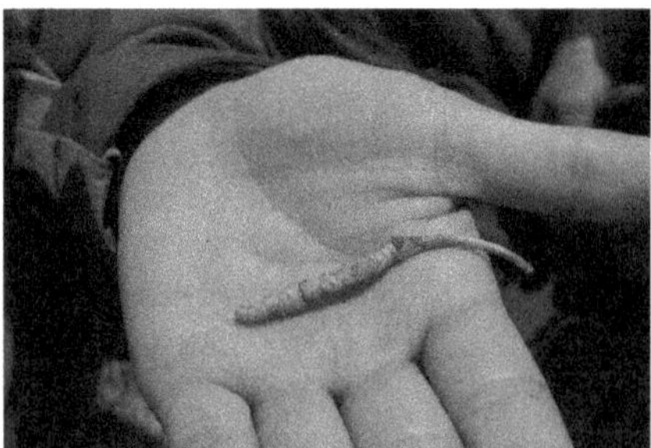

Figure 1.1a and Figure 1.1b Cordyceps sinensis

plants such as *Polygonum capitatum*, *Polygonum viviparum L.*, *Rhododendron parvifolium* and Humilis (Chen Tailu, 1973). Then, the bat moth larvae can be attacked at any time by Cordyceps.

When the spring season of the second year comes, the mycelium of Cordyceps grows above the ground and looks like a grass. In this way, the invaded larva shell and the Cordyceps mycelium together constitute a complete *Cordyceps sinensis*. The spores get nutrients through the worm's body and rapidly grow up. The worms are usually 4–5 cm, and the spores can grow to the length of the worms within one day. At this time, the worms are called 'the first-time harvested Cordyceps' (头草), and its medicinal quality is known as the most beneficial. As the spores grow up to about twice the size of the worm and form what is called 'the second-time harvested Cordyceps'(二草), the quality is considered to be less valuable, which is reflected in the lower price (Li, 2019).

The Origin of the Latin Name

A French scientist Dominicus Parennin went to China in 1723 to collect biological specimens, found *Cordyceps sinensis* and brought it back to Paris. From France a Cordyceps sample was taken to London by the British Levy as a rarity. In 1726, this Chinese precious medicinal material was exhibited for the first time at the Academy of Sciences in Paris, but the scientific community did not know what it was at the time (Shrestha, 2021). Japan's *New Agricultural News* recorded that Kawaguchi Ekai, a Japanese Buddhist monk, traveled to Tibet in 1728 and carried the Cordyceps specimens to Japan, and then *Cordyceps sinensis* was classified into the Clavaria (Tian, 2013). It was not until 1843 that English clergyman and mycologist Miles Joseph Berkeley discovered that the so-called *C. sinensis* was a kind of Ascomycota that parasitized larvae of the bat moth, and according to Chinese specimens it was classified into Sphaerodes and was named Spharia sinensis. In 1878, the Italian scholar Pier Andrea Saccardo officially named the species as Cordyceps, belonging to the fungi Dikarya, Ascomycota, Pezizomycotina, Sordariomycetes, Hypocreales, Ophiocordycipitaceae, Ophiocordyceps, Cordyceps, and this scientific name is still in use. The name *Cordyceps* is derived from the Latin words: cord (club) and ceps (head), and *sinensis* means Chinese. As a result of the classification of Cordycipitaceae and Clavicipitaceae families being updated in 2007, a group of Cordyceps species were classified into a new family Ophiocordycipitaceae (Sung et al., 2007).

General Characteristics

Fungi, as a biological class, are present in every ecosystem on Earth. Most varieties require mild conditions, but some can tolerate significant external stress (Singh Ahluwalia, 2021). In the case of *Cordyceps sinensis*, the fungus is dependent on the host insect, which is endemic to the Himalayan region—the area that makes up the natural boundaries of the Cordyceps' habitat.

Geographical Distribution

Cordyceps sinensis can only be found in the alpine meadow areas at an altitude above 3,000 meters and below the snow line in the Qinghai-Tibet Plateau and the surrounding areas. Therefore, only China, Bhutan, India and Nepal produce *Cordyceps sinensis*, 98% of which is produced in China. This area includes more than 100 counties in five provinces of China, including Qinghai, Tibet, Sichuan, Yunnan and Gansu. Geographically, the distribution of *Cordyceps sinensis* stretches from the Qilian Mountain in the north to the high mountain in the northwest of Yunnan in the south, east to the plateau mountains in Sichuan and westmost areas of the Himalayas, covering about one-tenth of China's land area.

The core production areas are Nagqu and Qamdo prefectures in the Tibet Autonomous Region and Yushu and Golog prefectures, Qinghai Province. The total output of the two places accounts for more than 80% of the national output. In addition, Cordyceps are harvested in Ganzi and the Aba Tibetan and the Qiang Autonomous Prefectures, Sichuan Province and in Deqin, Shangri-La, and Gansu prefectures, Yunnan Province.

The growth process of *Cordyceps sinensis* (Figure 1.2) comes in three stages:

(1) Ascospores of Cordyceps enter the worm's body and grow throughout the larva. The larva is still a worm in winter.
(2) During the second year's spring season, hyphae or mycelium in the worm's body starts to develop.
(3) In May to June, the mycelium takes over the worm's body, a rod-shaped fungal stroma grows out of the larva head, the top of the stroma expands and forms ascosporic spores in the ascus.

The abundance of Cordyceps depends on the number of host insects infected by spores and favorable external environmental conditions. That's why the

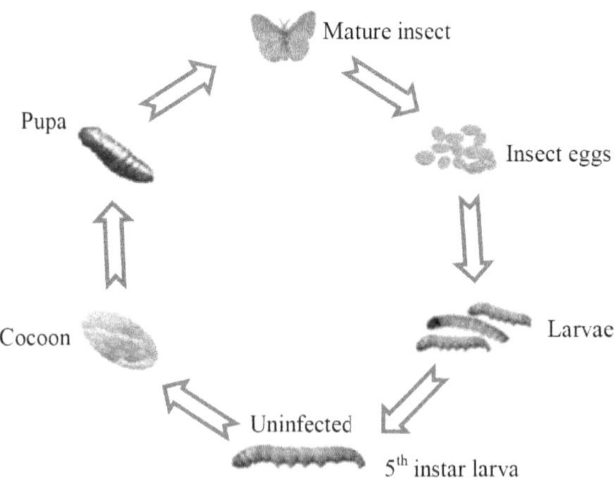

Figure 1.2 Growth cycle of Cordyceps bat moth

harvesting times vary. The Sichuan *Cordyceps sinensis* can be harvested from mid-May to the beginning of July in the solar calendar. The Yunnan *Cordyceps sinensis* begins to break through the surface of the wet soil after the first ten days of April and grows throughout late April to early May. At this time, the worm body attacked by the fungus is hard and full, while the stroma is short and tender. As the soil receives moisture from melting snow and the temperature rises, the stroma grows fast, but the worm body partially softens or rots. At the end of April, when the temperature rises to about 10°C in May, the stroma of the Tibetan *Cordyceps sinensis* begins to drill out of the ground, then it can be harvested after 10 days of growth. The best time for harvesting Cordyceps is from the middle of May to early June each year, when the snow in Qinghai plateau is melting and the temperature is gradually warming up. But in the Datong Mountain in Qinghai, harvesting Cordyceps can be extended to August or September due to the colder local climate (Wang, 2007).

What morphological characteristics (Table 1.1) are taken into account while harvesting Cordyceps depends on the growth cycle of the host insect (Figure 1.2), which can vary in different geographical areas due to local climatic variations.

Table 1.1 Morphological characteristics of Cordyceps

Morphological characteristics during growth	When the stroma of ascomycete grows out of the head of its host larva, it is solitary, slender as a baseball bat, 3–11 cm long. The shank is 3–8 cm in length, 1.5–4 mm in diameter. The upper part is the stroma, which is slightly enlarged, cylindrical, 1.5–4 cm long, brown, growing mostly out of asci shell, except for the small part of the apex. The asci shell is mostly trapped in the stroma. The apex which protrudes out of the stroma is oval, 250–500 mm long, 80–200 mm in diameter. Each asci shell has a plurality of long linear asci; each asci has eight ascospores with septum.
Morphological characteristics after removal from the soil	After mid to late June, *Cordyceps sinensis* enters into the spore emission stage. *Cordyceps sinensis* which is taken out of the soil is a dried worm body connected to the stroma, with a total length of 9–12 cm. The worm body is like a trimolter, which is 3–6 cm long and 0.4 0.7 cm thick. The worm's body is dark yellow, rough, with many transverse wrinkles on the back and 8 pairs of feet on the ventral surface; 4 pairs in the middle of the worm are clearly visible. Average weight of one dried Cordyceps strip depends on the regional variation and is usually within 0.14–0.35 g.
Morphological characteristics for medical production	Once it is established that the Cordyceps body is intact and it has no added object, which primary collectors might put in to increase the weight to get a better price, the next thing is to check the inside, to see if the section is full, white, slightly yellow, and the periphery dark yellow. The stroma that comes out of the head of the worm is rod-shaped and curved, and the upper part is slightly enlarged. The surface is taupe or dark brown, up to 4–8 cm long and 0.3 cm in diameter. When broken, the heart is empty and pink, slightly smelly and tasteless. The best Cordyceps has a bright yellow and plump worm's body, and short stroma.

Environmental Factors Affecting the Growth of Cordyceps

The natural environment of any species is 'the sum of those phenomena that enter a reaction system of the organism or otherwise directly impinge upon it to affect its mode of life at any time throughout its life cycle as ordered by any other condition of the organism that alters its environmental demands' (Mason & Langenheim, 1957). Since fungi are species that do not possess unique efficient biochemical or physiological mechanisms to adjust to their environments, they are highly dependent on their environment, but they are 'capable of becoming adapted to every condition of life' (Cooke, 1979). Unlike other fungi, *Cordyceps sinensis* is an endemic species, and its ability to adapt to noticeable changes in the environment is low. Furthermore, it is part of a longer ecological chain: Cordyceps relies on bat moths as host insect; the bat moth feeds on specific types of plants that require specific conditions in the environment to grow and reproduce. From that perspective, they can be used as a litmus test of the threshold of a local ecosystem.

Host Insect

Cordyceps sinensis is a complexus formed by Cordyceps infecting larvae of bat moths. In China, the major species of bat moth related to Cordyceps is Hepialus. According to the literature survey, more than 50 species of the Hepialus are found in China; a total of 20 species were found in Yunnan, 9 species were found in Qinghai, 14 species were found in Tibet, and 3 species were found in Gansu (Wang, 2007). In addition, there may be a small number of species of Hepialiscus, Forkalus and Bipectilus that can host *Cordyceps sinensis*.

Cordyceps bat moth is a completely metamorphic insect. Although the biological characteristics of different species are slightly different, they all go through four different periods; for example, eggs, larvae, pupas and adults. It takes about 4 years for the Cordyceps bat moth to complete a development cycle under natural environmental conditions, in which the larval stage lasts for more than 3 years. Larvae have the characteristics of generational alternation, hunger resistance, cold tolerance, drought and water tolerance, uneven distribution and omnivorousness.

The bat moth larva prefers low temperatures, and 40%–69% soil moisture provides the best condition for its growth. When the soil moisture is too low, the larva will spit out a large number of filaments to envelop the body and prevent evaporation of water to prolong its life. When the soil's moisture is too high, it is not in contact with the muddy water because of the dense, non-sticky particles fluffing on the body surface. The bat moth larvae production area is aggregated and distributed, and the density is up to 55 pieces/m^2; it is hard to find one every square meter or even every hundred square meters in the sparse place (Lu et al., 2002).

More than 50 species of bat moth have been reported so far, and the epicenter of its habitat in China is the Hengduan Mountains in the eastern part of the Qinghai-Tibet Plateau (latitude 27°–33° north latitude, 95°–103° east longitude), which accounts for 80% of the known species (Zhu & He, 2008). The bat moth also has a typical vertical distribution; the lower limit of its distribution is 3,000 meters above sea level in the southern boundaries of its habitat, and 2,500 meters above sea level in the northern border of its habitat; the upper limit is 5,100 meters above the sea level. The most suitable growth altitude is 3,600–4,800 meters. The most suitable growth soil is alpine meadow soil, which is concentrated on both sides of the round mountain and watershed on the slope of 15°–22° (ibid.).

Most important is the process of infection of bat moth by Cordyceps. Tibetan moth larvae are infected by Cordyceps in the soil, and the parasitic period coincides with the second molting of bat moth larvae (July to August). There are two channels for the infection. Either ascospores of Cordyceps adhere to the roots or soil of the plants, and when the larvae feed, they attack the body and infect them, or ascospores of Cordyceps adhere to larvae bodies that just exuviate or are damaged and penetrate under the skin (Hu et al., 2005).

The natural percentage of bat moth larvae infected by Cordyceps is 2.6%–16.1%, and the infection rate of 4–5 instar larvae is the highest, accounting for about 90% of the total infection rate. The 4–5 instar larvae that are in the stage of exuviating old skin, when the new epidermis just starts to grow, are most susceptible to infection. The 6 instar larvae are rarely infected, and the larvae below 3 instars are not infected, which is related to the activities, feeding and the strength of intracorporal antibacterial substances of different instar bat moth larvae (Zhou et al., 2013).

From July to August each year, Cordyceps begins to infect insects, and in October, the larvae die and become zombified. The stroma of Cordyceps grows out of the dead larva's head. From November to February, the stroma grows very slowly or even stops growing due to the low temperature and begins to resume growth in April. At the beginning of May, the stroma drills out of the ground, then after about 48 days, the ascospores mature and eject into the soil to infect the worm body and enter the next generation (ibid.).

Host Plant

The bat moth larva is an omnivorous insect that generally lives in the soil and dines on the tender roots of plants. There are many kinds of plants that can be eaten; they like the roots of the Polygonaceae plants like *Polygonum capitatum*, *Polygonum viviparum L.*, *Rheum pumilum*, etc. and also eat the roots of plants such as *Potentilla fruticosa*, Fabaceae and Astragalus, and Cyperaceae, Poaceae, Gentianaceae, Primulaceae and Juncaceae. In turn, these plants, especially *Polygonum capitatum* and *Potentilla fruticosa*, are used as traditional medicine to treat urinal and blood diseases (Lin et al., 2022; Chang, 1981). In addition, if no plants are available, ghost moth larvae can also eat humus in the soil under starvation (Zhang et al., 2015).

If the climate gets drier, then the area with mesic forbs, which are the key feed for bat moths, decreases (Chang, 1981). Increasing temperatures most likely will increase vegetative growth of alpine herbs such as *P. vivaparum*, but decrease the sexual reproductive growth of insects (Zhang & Tian, 2021). It is safe to assume that in the future there will be more variations in high meadow vegetation, and this will affect quantity and quality of wild Cordyceps.

Temperature

Low temperature is one of the main characteristics of the *Cordyceps sinensis* production area's climate. The average temperature in January is just below 0 °C; the extreme minimum temperature that was recorded hit below -20 °C, and the Nagqu at 4,507 m above sea level had a record of -41.2 °C. The average temperature in July is the highest but does not exceed 10 °C.

In the four regions (Tibet Nagqu, Tibet Linzhi, Qinghai Zadoi, Sichuan Kangding), the average monthly temperature is similar: from November to March it is around 0 °C or below, with April to October only a fraction warmer; therefore the *Cordyceps sinensis* habitat area is characterized by a long winter and short summer.

The growth and development of insects follows the cycle of effectively rising temperatures. Within the effective temperature range, the rate of development (the number of days required) is proportional to temperature: the higher the temperature, the faster the rate of development, and the fewer days required for development.

For the development of *Cordyceps sinensis*, host insects are the key factor, and they are also affected by the long winter season and the low average monthly temperature in the summer, so the effective temperature days for bat larvae development are short. This is the main reason why bat moth larvae need many years to complete their full development. The higher the altitude and the lower the average monthly temperature, the longer the period of development for larvae will be.

Humidity

Water is the main component for the successful development of *Cordyceps sinensis* and its host bat moth larvae, and for the optimal growth of the fungus, the soil moisture should exceed 80%. In fact, 80%–95% is the optimum atmospheric relative humidity for the growth and development of *Cordyceps sinensis*, when the stroma grows fast and becomes hypertrophic. On the contrary, air humidity less than 70% is not conducive to the growth of *Cordyceps sinensis*, when the stroma grows extremely slowly or becomes dry, and the development is disrupted. In addition, humidity also has an important influence on the development of the bat moth, especially of its larvae.

The amount of precipitation seriously affects the growth and yield of *Cordyceps sinensis*. According to the reports, in Suo County, Biru County, Baqing County and Jiali County in the eastern part of the Nagqu, which are the main production areas of *Cordyceps sinensis*, the precipitation (250–500 mm per year) is significantly higher than the precipitation of Nierong County and Nagqu County (200 mm per year), which are areas that produce less Cordyceps. The amount of snowfall in early spring directly affects the production of *Cordyceps sinensis*: the more snowfall, the higher the yield of *Cordyceps sinensis* in the coming harvesting season; without snowfall there will be little yield (Zhang et al., 2011).

As mentioned, the growth and development of bat moth larvae also requires a specific level of humidity. June to September is the best period for the growth and development of meadow plants. The temperature is rising, there is sufficient sunshine, abundant rain, and the bat moth larvae have plenty of food and can rapidly grow.

Humidity has an important influence on adult bat moths and their natural enemies. The humidity suitable for adult courting, mating and spawning is 75%–85%. Rainfall can directly affect the activity of adults (courtship, mating), so that effective egg production is reduced, and the population of the next generation is affected as a result. At the same time, the high-humidity soil is conducive to the germination, infestation, pathogenicity and growth of the spores of pathogens that attack larvae, resulting in increased mortality of the bat moth. The newly hatched larvae have poor mobility, and the rain during the incubation period has a direct lethal effect. As a result, it might mean fewer Cordyceps being produced.

Light

Light has a regulatory effect on the growth and second metabolism of fungi, and Cordyceps are no exception (Dong et al., 2012). The Qinghai-Tibet Plateau is an area rich in sunlight, and the annual sunshine hours and annual sunshine percentages of the *Cordyceps sinensis* areas in the middle and eastern parts of the plateau and southeastern Tibet are 2,500 h and 60% respectively (Zhang et al., 2011). Studies have shown that light has an effect on the growth and development of Cordyceps stroma and ascospores (ibid.).

Lack of light can inhibit the normal growth and development of Cordyceps stroma and ascus. Under dark conditions, after it has broken through the soil, Cordyceps stroma grows at an average speed of 0.4 mm/d; the growth rate decreases to 0.06 mm/d after 36 d, and then stops growth after 70 d generally, while it stops generating asci and ascospores.

Light intensity and illumination time affect the external morphology and normal growth of *Cordyceps sinensis*. If the light is strong and the exposure is long, the growth of the stroma is suppressed, but the stroma is thick. On the contrary, if the light is weak and the exposure is short, the stroma is excessively long and slender.

Strong ultraviolet radiation can inhibit the excessive length of the stroma and increase the germination rate of ascospores. The average height of *Cordyceps sinensis* growing under strong sunlight at 4,000 m above sea level is 40.1 mm, and the average height of *Cordyceps sinensis* growing under a glass cover is 42.5 mm. The germination rate of ascospores after ultraviolet irradiation is higher than with other treatment methods.

The effect of sunlight on bat moths mainly manifests in two aspects. First, the light can increase the temperature of the atmosphere and soil, enhance the vitality of bat moth larvae in the soil, increase the metabolism level of the worm body, and promote the growth and development of bat moths in different states. Secondly, the increase in sunlight and ambient temperature promotes photosynthesis and growth of plants, and provides a richer food resource for bat moth larvae (Zhang et al., 2011).

For artificial production of *Cordyceps militaris*, it was found that red- and pink-colored light can increase the contents of bioactive components, such as carotenoids, adenosine and cordycepine (Dong et al., 2012).

Altitude

The average elevation of the Qinghai-Tibet Plateau is above 4000 m, and such a high-altitude location is best for the growth of *Cordyceps sinensis* and its host bat moth. Although the existing studies have proved that the *Cordyceps sinensis* distributed in different production areas of the Qinghai-Tibet Plateau belong to the same species, varying altitudes affect the ability of Cordyceps to infect moth larvae (Zhang et al., 2011).

It can be seen from Table 1.2 that the altitude of the distribution range of *Cordyceps sinensis* in the two core Cordyceps habitat areas of Nagqu, Tibet and Yushu, Qinghai is higher than for Sichuan, Yunnan and Gansu, which are marginal distribution areas. At the same time it can be found that the lower limit of the altitude is moving up and the distribution range is shrinking. The increase of the lower limit of the altitude of the distribution range of *Cordyceps sinensis* depends on other environmental factors such as temperature, soil moisture, etc., and these factors are affected by natural environmental conditions and human disturbance to varying degrees.

Mountain Slope: Direction and Gradient

Cordyceps are most commonly found at sunny slopes of 15°–60°, and especially slopes at 15–25°. Cordyceps mainly grow on the windward side of a mountain and rarely grow on the leeward side. If the mountain gradient is too large, Cordyceps rarely grow there. The amount of distribution decreases correspondingly with the

Table 1.2 Historical altitudes of areas that produce *Cordyceps sinensis*

Area	*Distribution range (m)*	*Optimum altitude range (m)*	*Upper limit (m)*	*Lower limit (m)*
Nagqu, Tibet	4,100–5,000	4,300–4,800	4,100	5,000
Yushu, Qinghai (in the 1960s)	3,500–5100	4,000–4,600	3,500	5,100
Yushu, Qinghai (in the 1990s)	4,100–5000	4,300–4800	4,100	5,000
Northwest of Yunnan (in the 1960s)	3,600–5000	4,000–4,600	3,600	5,000
Northwest of Yunnan (in the 1980s)	4,000–5,000	4,000–4,600	4,000	5,000
Lixian, Sichuan	3,500–4,700	4,400–4,600	3,500	4,700
Kangding, Sichuan	3,200–4,700	3,650–4,250	3,200	4,700
Sunan, Gansu	3,350–4,250	3,800–4,100	3,350	4,250

Note: Data from Zhang et al. (2011). Factors influencing the occurrence of Ophio*Cordyceps sinensis*. *Acta Ecologica Sinica*, *31*(14), 4117–4125.

increase of the gradient of the mountain, and Cordyceps grow especially less in sloped areas exceeding 50°. In shady slopes with flat grasslands or swamps, fewer Cordyceps can be found (Zhang et al., 2011).

Vegetation and Soil

Cordyceps are found in alpine shrub meadows, alpine meadows and stratified forest belts. *Picea asperata*, *Quercus acutissima*, etc. grow on the upper layer of the stratified forest belt, and shrubs such as *Quercus aliena*, *Camellia oleifera*, *Dalbergia hupeana*, *Rhus chinensis* and Rhododendron grow on the upper layer. At altitudes of 3,000–3,500 m, the most common plants are *Rhododendron vernicosum* and *Polygonum macrophyllum*; at 4,000–4,500 m, mostly *Rhododendron araiophyllum*, *Polygonum macrophyllum* and Asteraceae grow; at more than 4,500 m, Polygonaceae plants are the dominant varieties (Zhou et al., 2013).

The soil where the Cordyceps is found is generally fine-grained; the texture is sandy loam or light loam. The soil known as plateau meadow soil has a pH of 4–6 and is the leading soil type of the main production areas of Cordyceps. In addition, there are yellow or brown soils located at higher altitudes; the humus layer is dark brown and the soil is neutral or slightly alkaline, which can produce Cordyceps. In general, the humus layer of soil in areas that produce Cordyceps is thicker, permeability and drainage is higher, the soil is mostly acidic, and the organic matter content is up to 8%–20% (ibid.).

Other Factors to Consider

In addition to the aforementioned factors, survival of Cordyceps species depends on the amount of feed for its natural host—the bat moth and its natural enemies, such as rodents (pikas), birds, arachnids and pathogenic microorganisms.

Overharvesting of Cordyceps and overgrazing directly affect the growth of meadow plants and reduce the biomass growth of aboveground and underground organisms, leading to the degradation of meadows and the decline of ecological functions, thus changing the environment of *Cordyceps sinensis*, which affects its growth. Current predatory digging has gradually become the major factor affecting the production of *Cordyceps sinensis*.

In the 1950s, the production of Cordyceps in China reached more than 1 million kg. After this period of overharvesting, the annual production dropped to about 500,000 kg in the 1960s. By the 1990s, when demand and prices went up, natural production was only about 300,000 kg per year. In the 21st century, the produced amount has fallen further (Zhang & Tian, 2021).

Conclusion

To sum up, Cordyceps species *C. sinensis* is a parasitic heterotroph, which is endemic only to the Himalayan region. It has been known for its beneficial qualities since the 8th century AD. The Western world discovered this species only in the 18th–19th centuries, and in many aspects the fungus is still enigmatic to contemporary science. Active scientific research is still under way to bring light to the biology, morphology and biochemistry of Cordyceps. Considering the character of the species—its rarity and high sensitivity to major changes in its environmental complex—a detailed analysis of variables of the ecosystem provides a foundation for future successful conservation programs.

References

Borman, F.H., & Likens, G.E. (1970). The nutrient cycles of an ecosystem. *Scientific American*, October, 92–101.

Buddhist Digital Resource Centre. (2006). Yue wang yao zhen: Electronic reproduction. https://archive.org/details/bdrc-W30387/page/n11/mode/2up

Chang, D.H.S. (1981). The vegetation zonation of the Tibetan Plateau. *Mountain Research and Development*, *1*(1), 29–48.

Chen Tailu, T.C.M.C. (1973). A preliminary study on the biology of the 'Insect Herb, Hepialus Armoricanus Oberthür.' *Acta Entomologica Sinica*, 2, 198–202.

Cooke, W.B. (1979). *The Ecology of Fungi*. CRC Press: N.Y.

Dong, J.C., Lei, C., Zheng, X., Ai, X.R., Wang, Y., & Wang, Q. (2012). Light wavelengths regulate growth and active components of Cordyceps militaris fruit bodies. *Journal of Food Biochemistry*, *37*(5), 578–584.

Holling, C.S. (1973). Resilience and stability of ecological systems. *Annual Review of Ecology and Systematics*, 4, 1–23.

Hu, Q-X., Liao, C-Z., & Wang, X. (2005). Countermeasures for protecting, developing and utilizing Chinese Caterpillar Fungus and their resources in China. *Chinese Journal of Agricultural Resources and Regional Planning*, 5, 47–51.

Li, Z. (2019). Qinghai Province Guoluo characteristic economy develops investigation. *Qinghai Financial*, 9, 36–39.

Lin, Y., He, L., Chen, X., Zhang, X., Yan, X., Tu, B., Zeng, Z., & Hu, M. (2022). Polygonum capitatum, the Hmong medicinal flora: A comprehensive review of its phytochemical, pharmacological and pharmacokinetic characteristics. *Molecules*, *27*(19), 6407.

Lu, L., & Liu, S. et al. (2002). New progress of *Cordyceps sinensis* research. *Bulletin of Biology*, 6, 4–6.

Mason, H.L., & Langenheim, J.H. (1957). Language analysis and the concept 'environment.' *Ecology*, *38*(2), 327–340.

Shrestha, B. (2021). Yarsa Gumba (Ophio*Cordyceps sinensis*): A national pride of Nepal. https://www.researchgate.net/publication/348560322_Yarsa_Gumba_Ophiocordyceps_sinensis_A_national_pride_of_Nepal

Singh Ahluwalia, A. (2021). Foreword. In A.M. Abdel-Azeem, A.N. Yadav., N. Yadav, & Z. Usmani (Eds.), *Industrially important fungi for sustainable development*. Springer: Cham.

Sung, G.H., Hywel-Jones, N.L., Sung, J.M., Luangsa-Ard, J., Shesta, B., & Spatafora, J.W. (2007). Phylogenetic classification of Cordyceps and the Clavicipitaceous fungi. *Studies in Mycology*, 57, 5–59.

Sutherland, W.J. (2022). *Transforming conservation: A practical guide to evidence and decision making*. Open Book Publishers: Cambridge. https://doi.org/10.11647/OBP.0321

Tian, X. (2013). *Artificial cultivation of OphioCordyceps sinensis phorozoon and its effect on HepG2 cell apoptosis* [doctoral dissertation]. China Agricultural University, 94.

Wang, L. (2007). Investigation on the resource distribution of *Cordyceps sinensis*. *Forestry of Gansu*, 4, 40–41.

World Mushroom Society (WMS). (2022). Ultimate guide to Cordyceps mushrooms. https://worldmushroomsociety.com/cordyceps-mushroom-guide

Xinhua. (2021). Remarks by Chinese President Xi Jinping at Leaders' Summit on Climate, 4 April 2021. www.xinhuanet.com/english/2021-04/22/c_139899289_2.htm

YiLanTingMuChan. (2022). *Cordyceps sinensis.*
https://baijiahao.baidu.com/s?id=1733706176314090981&wfr=spider&for=pc

Yue, K., & Ye, M. (2013). The genus Cordyceps: A chemical and pharmacological review. *Journal of Pharmacy & Pharmacology*, *65*(4), 474–493.

Zhang, C., & Tian, Z. (2021). Analysis of the development status and sustainable utilization of *Cordyceps sinensis* resources. *China Edible Mushroom*, *40*(10), 79–88.

Zhang, G., Yu, J.F., Wu, G.G., & Liu, X. (2011). Factors influencing the occurrence of Ophio*Cordyceps sinensis*. *Acta Ecologica Sinica*, *31*(14), 4117–4125.

Zhang, Z., Liu, X., Xu, H., & Li, Y. (2015). Correlation between the yield of Cordyceps and host larvae. *Qinghai Journal of Animal Science and Veterinary Medicine*, 5, 27–29.

Zhou, Z., Hong, D., Niu, Y., Li, G., Nie, Z-L., Wen, J., & Sun, H. (2013). Phylogenetic and biogeographic analyses of the Sino-Himalayan endemic genus Cyananthus (Campanulaceae) and implications for the evolution of its sexual system. *Molecular Phylogenetics and Evolution*, 68, 482–497.

Zhu, D., & He, R. et al. (2008). Ecological habits of bat moth on Qinghai-Tibet Plateau of China. *Xizang Agricultural Science and Technology*, 2, 16–18.

2 Medicinal Value and Use of Cordyceps

Jiping Sheng, Yiyuan Miao, Yiyun Wen, Yijing Xin, Shengming Shen, and Ksenia Gerasimova

Introduction

Traditional Chinese Medicine (TCM) is the quintessence of Chinese science based on the knowledge empirically collected by Chinese scholars and medics over centuries. For Western medicine, it is still often unknown territory but serves as a source of inspiration to conduct new research, especially on its most cherished treatments, including Cordyceps. Westerners consider TCM 'an astonishing phenomenon' where 'an unproven and scientifically unverifiable set of ancient and foreign healing methods [has] been able to arouse such great interest among Western industrial nations, to the point where it must be considered a serious component of their health care systems' (Unschuld, 2018, p. 2). Modern China has three types of medicine: traditional Chinese, Western, and 'the integrated one' (Keji & Hao, 2003). Cordyceps has led to the merging of Western and traditional Chinese medical sciences, and there are more untapped opportunities for future research and development of innovative treatments for serious diseases, such as cancer and viral infections. For many years, Cordyceps was an 'over-the-counter medicine/tonic' advertised as a Chinese herb with anti-aging, 'pro-sexual,' anti-cancer and immune-boosting effects, although with 'poor supporting scientific evidence.' Consumers used to believe in the potency of *C. sinensis* mostly because of the TCM brand name (Paterson, 2008). The COVID-19 epidemic marked the introduction of academic research to understand the unique biochemical characteristics of Cordyceps and to develop new medications. In this new landscape, demand for Cordyceps and its derivative products will likely keep increasing.

Ancient Medical Literature on Cordyceps

The first records of TCM are dated to the era of Huangdi or the Yellow Emperor, during the period 2698–2598 BC. The best known medical ancient Chinese textbook is the *Huang-ti Nei Ching* (Yellow Emperor's Canon of Internal Medicine). In one way or another, most Chinese medicine is based on the Yellow Emperor's Code (Catic et al., 2018).

Historians, having studied different manuscripts, argue that TCM is 'the oldest continuously practiced, scientific medical system in the world,' and that it should

DOI: 10.4324/9781003429753-4

not be equated to shamanism or quackery, but should be considered as a complex and precise health care system based on Chinese civilization (ibid.). TCM has structured principles based on the importance of the environment to human health, borrowed from the Taoist philosophy. The Yellow Emperor's Canon explains the correlation of two opposite aspects within one phenomenon, the famous yin-yang model, balancing under the influence of the five elements (water, fire, metal, wood, and earth) (Chen Marshall, 2020). A patient's diagnosis was defined by measuring their pulse, and the treatment aimed to 'guide [the] patient towards Tao' and the nutritional value of plants and minerals (Curran, 2008).The traditional Chinese medicinal texts, also commonly referred as the Bencao texts, have been circulated since the Warring States period (481 BCE–221 BCE) and are still widely cited. Since these Chinese Materia Medica are translated as 'rooted in herbs,' they indeed prescribe plant-based medicines (Liang & Zhao, 2017).

According to our approximate review, there were more than 70 traditional Chinese medicine texts describing the efficacy of *Cordyceps sinensis* in ancient China. In general, Cordyceps are presented as able to nourish yin and yang, treat a continuous cough and supplement the human body against all kinds of health damage. Cordyceps treatment has the same efficacy as ginseng and velvet antler, with the three of them forming the famous triage of TCM. According to ancient tractates, Cordyceps are suitable for everyone: young and old, healthy and sick.

Table 2.1 Ancient medicinal texts on Cordyceps

Resource	*Content*	*Reference*
Wang Ang's study, titled *Bu Tu Ben Cao Bei Yao* (《补图本草备要》)	Written in 1694 AD, it states that *Cordyceps sinensis* is 'sweet in flavor and warm in nature' and has the effect of reducing general health deficiency. It strengthens vital energy and improves the strength of kidneys and lungs, and thus was suitable for patients with tuberculosis, a cough, hemoptysis, dyspnea, and night sweats.	Wang, 2022
Ben Cao Gang Mu Shi YI (《本草纲目拾遗》)	*Cordyceps sinensis*, warm, can supply vital essence and protect lungs. Drinking a few Cordyceps soaked in wine can cure pain in the waist and knees and benefit the kidneys. Cooking it with the male duck is suitable for the elderly . . . can cure lots of damage.	Zhang, 1871
Ben Cao Cong Xin (《本草从新》)	*Cordyceps sinensis* has the medicinal effect of 'preserving lung and kidney, stopping bleeding and removing phlegm.'	Wu, 2013
Liu Ya Wai Bian (《柳崖外编》)	Eating stewed duck meat with Cordyceps can notably benefit the body.	Zhuan, 1969
Gan Yuan Xiao Shi (《柑园小识》)	For those who are ill-healed and debilitated after illness, taking a duck with Cordyceps inside is equal to taking 1 Jin (50g) ginseng.	Zhu, 1767
Ben Cao Zheng Yi (《本草正义》)	Cordyceps, which can tonify the kidney and strengthen the yang, is suitable for han but not for xu re and can cure the spleen and kidneys.[1]	Zhang, 1920

For a Western audience, Traditional Chinese Medicine (TCM) and Tibetan Medicine (TM) can be mixed up with each other. The truth is that they are rather different yet still related. When Wencheng, a royal princess of the Tang dynasty, married a Tibetan king, the dowry also included 'a hundred prescriptions for the treatment of four hundred and four diseases, five kinds of diagnostic methods,' and Chinese doctors were invited to Tibetan courts. The key foundation for TM is the flora of the Tibetan Plateau. Thus, Cordyceps are included among the herbal treatments that are shared by TCM and TM (Zhao et al., 2019).

In the past, Tibetans collected wild plants and mushrooms for personal use, but now in the 21st century, they have started to collect them mostly for sale, and not so much for personal treatment, since the demand for traditional medicinal remedies has substantially increased (Boesi, 2007). Unfortunately, the projections are such that Tibetan traditional knowledge of the plant world, particularly at the popular level, is going to disappear in the near future, as several other aspects of the Tibetan traditional lifestyle have already been lost (ibid.; Boesi, 2014).

Chemical Composition

Cordyceps sinensis contains several microelements that make them so valuable, such as polysaccharide, Cordycepic acid, adenosine, amino acids, vitamins, and active peptides.

Cordycepic Acid (D-mannitol)

In the 1950s, Chatterjee, Srinivasan, and Maiti (1957) reported the isolation of a new compound from the *Cordyceps sinensis*, which was named 'cordycepic acid' (Sprecher & Sprinson, 1963). Later it was found that 'cordycepic acid' is neither a new compound nor an acid, but in fact, this white chrystallic powder is D-mannitol, D-enantiomer of mannitol (Xu et al., 2023). Chemically, mannitol is a polyol; it is similar to xylitol and sorbitol. It has been used in the food production and pharmaceutical industry, as it is an osmotic diuretic, a sweetening agent, an antiglaucoma drug, a metabolite, and a food agent (NTP, 1992). D-mannitol can increase oxygen in the blood, stabilize blood pressure, and reduce cerebrovascular accidents, thus it is useful for cardiovascular concerns. It has shown to have antimicrobial and antiviral effects. It can be used to reduce pigmentation and skin irritation. D-mannitol has also been used as an infertility treatment in men and women (Turliuc, 2019).

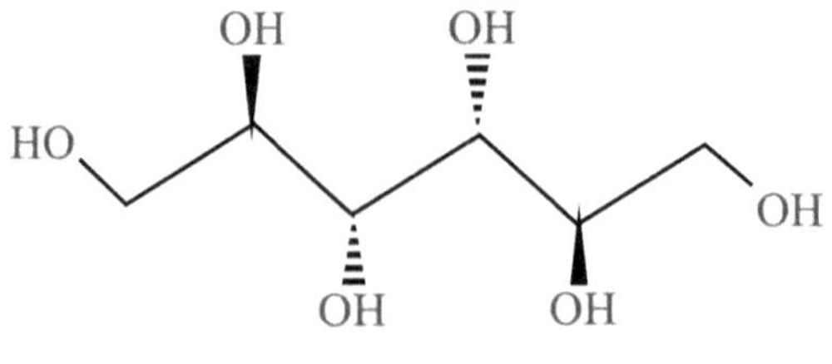

Figure 2.1 Chemical structure of Cordycepic acid

Nucleic Acid-Related Compounds

Cordycepin (Figure 2.2), which is also known as 3′-deoxyadenosine, is a derivative of adenosine, which inhibits the release of neurotransmitters in the central nervous system (Xu et al., 2023). It is an endogenous purine nucleoside, which has an impact on diastolic blood vessels and can lower blood pressure, reduce heart rate, stimulate adrenaline, stimulate the production of steroid hormones, inhibit platelet aggregation, relax vascular smooth muscle, and has an anti-convulsion and anti-tumor effect (Paterson, 2008). It has obvious inhibitory effects on Bacillus subtilis, Mycobacterium tuberculosis avium, and Ehrlich ascites cancer cells (Huang et al., 2019).

Other Bio-active Components (Amino Acids, Vitamins, Minerals, Peptides)

Cordyceps sinensis powder is rich in amino acids, containing 18 kinds, including essential amino acids, which strengthen one's body and enhance immune function.

Cordyceps contains vitamins (Vit E, K, and water-soluble vitamins B1, B2, and B12) and a variety of elements (K, Na, Ca, Mg, Fe, Cu, Mn, Zn, Pi, Se, Al, Si, Ni, Sr, Ti, Cr, Ga, V, and Zr) (Zhu et al., 1998). Aside from essential amino acids, *Cordyceps sinensis* contains proteins, peptides, polyamines, some uncommon cyclic dipeptides and polyamines (Holliday et al., 2010). It was found that Cordyceps can produce myriocin (C21H39NO6) and a complex family of non-ribosomal polypeptides, each containing two α-aminoisobutyric acid residues, which have shown to have anti-tumor effects (Xu et al., 2023).

Cordyceps' polysaccharides are a macromolecular active ingredient, which activates the reticuloendothelial system (RES) and peritoneal macrophages (PM) (Cheung et al., 2009). Polysaccharides mainly play a regulatory role in the immune system by activating these cells, which promotes cytokine production. Cordyceps polysaccharides can be used in anti-cancer, anti-diabetes and anti-atherosclerosis therapies and help protect the liver, intestines, and kidneys (Yuan, 2020).

Cordyceps can also be used to produce intracellular selenium-enriched polysaccharides (ISPS). Experiments have shown that ISPS of *C. sinensis* can reduce

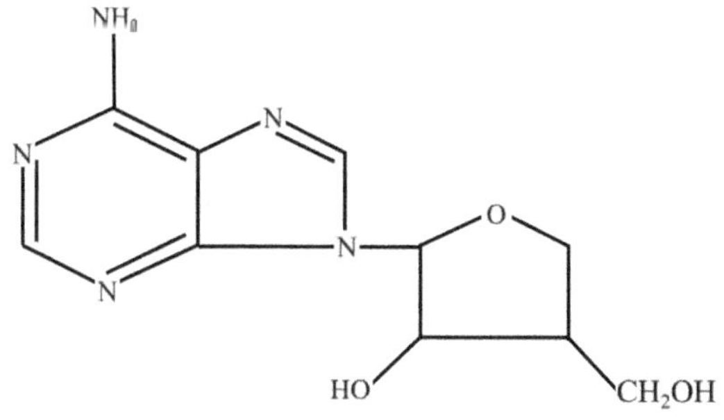

Figure 2.2 Chemical structure of Cordycepin

blood glucose levels and improve the antioxidant capacity of rats with diabetes induced by streptozotocin (Yang & Zhang, 2016).

Modern Medical Literature on Cordyceps' Medicinal Use

Cordyceps-based therapy might be effective in treating several global killer diseases, including cardiovascular pathologies and cancer. Below is a summary of some studies that prove the advanced pharmacological value of Cordyceps extracts.

Pharmacological Effects and the Evidence

Impact on the Immune Function

Different medications from *Cordyceps sinensis* can significantly stimulate the phagocytic function of the mononuclear-phagocytic system, increasing activity of alkaline phosphatase, enhancing the phagocytic capacity of hepatocytes, and promoting the proliferation of spleen macrophages (Shin et al., 2003). It can also promote protein metabolism. Thus, it is a non-specific immune enhancer that improves the body's immunologic function without toxic side effects on normal cells (Zhang et al., 2011).

Results from the research on anti-corneal transplantation rejection suggest that *Cordyceps sinensis* can play a role as an immunosuppressive agent after corneal transplantation. It selectively acts on T suppressor cells in T lymphocyte subsets to exert cellular immune regulation (Zhu & Yu, 1990). The water extract of Cordyceps and Cordyceps fungi has obvious protective effects on the thymus of mice and animals with low spleen T cells function, and its application can increase the number of peripheral white blood cells of normal mice (Chen et al., 1991).

From the perspective of regulating humoral immunity, the oral administration of Cordyceps to mice has increased the level of serum hemolysin. In vitro experiments can directly stimulate the proliferation of mouse thymocytes, and the stimulation is dose-related. Lab mice experiments have shown that *Cordyceps sinensis* extract has a multi-faceted immune function or two-way regulation: it enhances or inhibits the function of different lymphocyte subsets, which does not affect the function of the body's hematopoietic system and has no lymphocyte toxicity (Hao et al., 2012).

Anti-Cancer Effect

The intraperitoneal injection of natural *Cordyceps sinensis* and artificial Cordyceps mycelium water extracts significantly inhibited the primary growth and spontaneous lung metastasis of subcutaneously transplanted Lewis lung cancer in mice (Khan et al., 2010). The aqueous extract of *Cordyceps sinensis* significantly enhanced the ability of natural killer cells to kill tumors in active leukemia patients but inhibited the activity of LAK cells. It opened a discussion to explore the function of enhancing natural killer cells' activity by *Cordyceps sinensis* water extract to treat leukemia patients (Xu et al., 1992). In addition, the Cordyceps alcohol extract has reduced the incidence of cancer development in mice. It enhanced the activity of natural killer cells in vivo and in vitro studies on mice and humans. Thus, it was shown that Cordyceps extract inhibits the formation of mouse tumors (Wang et al., 2021).

Cordyceps polysaccharide (PSCS) has an effect on the proliferation and differentiation of human leukemia cells, as it can promote the activity of blood mononuclear cells (PSCS-MNC-CM) and significantly inhibit the proliferation of U937 cells, inhibiting growth rate by 78%–83%. PSCS-MNC-CM treatment differentiated approximately 50% of cells into mature mononuclear leukocytes, which are capable of expressing non-specific esterase activity (NSE) and surface antigens CD11b, CD14 and CD68. The levels of IFN-Y, TNF-a and IL-I in normal MNC-CM were very low, and their content was significantly increased under the stimulation of PSCS (Chen et al., 1997).

Effect on the Cardiovascular System

Subcutaneous injection of Cordyceps alcohol extract can significantly prolong the survival time of mice by increasing the nutritional blood flow of the heart and brain tissue under both normal pressure and hypoxic conditions (Ye, 2017). Intravenous injection of Cordyceps alcohol extract and fermentation broth can prevent myocardial ischemia in lab animals caused by pituitrin (Zhu et al., 1998).

Intravenous injection of the extract of *Cordyceps sinensis* into mice can significantly inhibit aortic pressure. Studies have shown that protein extract-induced vasodilation is indirectly caused by nitrogen-containing compounds released by the endothelial cell layer and endothelial cell layer-derived super-factors, which contain components that can relax the vessel in order to reduce aortic pressure (Chen, 1995).

Regulates Body Metabolism

Experimental studies on blood glucose regulation showed that *Cordyceps sinensis* has a good hypoglycemic effect, and the effect shows a dose-effect relationship. Subcutaneous injection of *Cordyceps sinensis* alcohol extract and fermented extract can significantly reduce serum cholesterol and triglyceride levels in hyperlipidemia mice, and also reduce serum cholesterol levels in normal mice (Yu et al., 2015). Another study shows that intragastric administration with *Cordyceps sinensis*-based medication can significantly reduce plasma triglyceride, total cholesterol, low-density lipoprotein cholesterol and very low-density lipoprotein cholesterol while high-density lipoprotein cholesterol is significantly increased, thereby improving arteriosclerosis. The mechanism of hypolipidemic effect is such that Cordyceps extract can activate lipoprotein esterase in the capillary wall and extravascular tissues (mainly adipose tissue), increase its activity and enhance the decomposition of triglycerides (Koh et al., 2003).

Sedative Effect

Intraperitoneal injection of Cordyceps and a Cordyceps decoctum can significantly reduce spontaneous activity and prolong sleep time in mice. The effect of the decoctum is stronger than an injection, both of which can significantly prolong the sleeping time of mice induced by pentobarbital sodium, which produces a sedative effect. Cordyceps alcohol extract also has an obvious sedative effect, but there is no disappearance of the righting reflex at the dose close to poisoning, so it is considered that there is no hypnotic effect (Sun et al., 2017).

Table 2.2 Clinical use of Cordyceps-based medicine

1) **Treatment of sexual dysfunction**
   Yang Wenzhi et al. (1985) used Cordyceps (1g each time, 3 times per day) to treat sexual dysfunction; the effective rate of the patients taking processed Cordyceps was 64.15%, while for those taking raw Cordyceps it was 31.57%.
2) **Treatment of hyperlipidemia**
   Wu Bingying et al. (2007) observed that the artificial cultivation of Cordyceps plays a certain role in reducing γ-GT and cholesterol, and has an anti-fatigue effect.
3) **Treatment of thrombocytopenia**
   Chen Daoming et al. (1995) used Cordyceps capsules to treat primary thrombocytopenia with an effective rate of 83.3%.
4) **Therapeutic effect on coronary heart disease and arrhythmia**
   Yang Chaokuan et al. (1990) used Ningxinbao capsule (oral 500mg) for the treatment of 54 patients with arrhythmia. ECG monitoring results showed a total effective rate of 81%; it had the effect of speeding up conduction, regulating heart rate, and improving heart function.
5) **Effects on chronic hepatitis, hepatitis B, and cirrhosis**
   Liang Huijing et al. (2006) used oral *Cordyceps sinensis* and hepatitis B surface antibody-positive placenta to treat chronic hepatitis B and achieved ideal results. Through the clinical observations of Wang Yunong et al., Cordyceps were seen to increase CD4+, NK cells, and the ratio of CD4+/CD3+, indicating the regulatory function of the cellular immune function of patients with chronic hepatitis B. Liu Cheng et al. used artificial Cordyceps mycelium preparations to treat patients with liver cirrhosis, and liver function was seen to improve to varying degrees.
6) **Treatment of tumors**
   Cordyceps can be used as an adjuvant treatment for malignant tumors. Zhang Jinchuan et al. (1986) made capsules of Cordyceps for treatment, and 93% of patients showed improved clinical symptoms.

Anti-COVID-19 Cordyceps-Based Treatment

Traditional Chinese Medicine proved to be the essential supplementary tool in combating the epidemic. COVID-19 has become a historic event of the 21st century, having pushed governments to take unprecedented measures. Globally, as of 5 January 2023, there have been 657,430,133 confirmed cases of COVID-19, including 6,676,645 deaths, reported by WHO. Up to 21 December 2022, a total of 13,073,712,554 vaccine doses had been administered (WHO, 2023).

The outbreak of Coronavirus Disease 2019 (COVID-19) began in the People's Republic of China (PRC) and globally led to disruptive patterns in the supply chain of personal protective equipment and antibiotics and the critical dependence of the USA and other countries on supply chains based in China (CRS, 2020). China, the global drug producer, also had episodes of shortages of antiviral and anti-fever medicines during the pandemic (Pierson et al., 2022). But it was less acute, as traditional Chinese herbs were available. Vaccination was rather well-administered, as the two main vaccines manufactured in China, CoronaVac and Sinopharm, were given to about 90% of the Chinese population (Stevenson, 2022). Meanwhile, TCM has also proven to be helpful alongside the vaccination process. As the human body needed time to produce antibodies after the vaccine dose was processed, TCM-based medicines were providing immediate protection, although it was not as prolonged as the vaccine (Lee, 2022).

From the beginning of the outbreak in Wuhan in December 2019 up till February 2020, in 85% of the total reported mild cases across the various provinces in China, patients were treated with TCM, improving their recovery by two days and reducing lung inflammation by 22% in comparison to average recovery rates and provided early preventive treatment as well (Wu et al., 2020). While Cordyceps were not on the official list of approved medications to treat COVID-19, and the most effective way to treat the virus is to be vaccinated, the interest in its biochemical qualities has grown significantly. Chinese patients—at least those who could afford its high price—used Cordyceps as an immune-boosting treatment during the pandemic. Bioactive components of Cordyceps, such as ergosterol and cordycepin, became the target of new research during the pandemic in order to develop new medicinal products (Yoneyama et al., 2022; Verma, 2022).

The accumulated data from traditional texts and most recent research shows the medicinal qualities of *Cordyceps sinensis* are particularly valuable for treatment of widespread global killer diseases and serious symptoms. During the COVID-19 pandemic, the interest in Cordyceps and its derivatives was renewed, and most likely academic and corporate research will continue to discover useful clinical applications for Cordyceps.

Although Cordyceps has a wide range of benefits, it is by no means a panacea for aforementioned diseases. Cordyceps has low toxicity and no obvious abnormalities in long-term toxicity tests (Long et al., 2021), but it is advisable to consult with a doctor and not self-prescribe it.

Note

1. Han (cold), xu (heat) and re (deficiency) are the three symptoms described in TCM.

References

Boesi, A. (2007). The nature of Tibetan plant nomenclature. *The Tibet Journal*, *XXXII*(1), 3–28.

Boesi, A. (2014). Traditional knowledge of wild food plants in a few Tibetan communities. *Journal of Ethnobiology and Ethnomedicine*, 10, 75.

Catic, R., Oborovic, I., Redzic, E., Sukalo, A., Skrbo, A., & Masic, I. (2018). Traditional Chinese medicine—An overview. *International Journal on Biomedicine and Healthcare, 6(*1), 35–50. doi: 10.5455/ijbh.2018.6.35–50

Chatterjee, R., Srinivasan, K.S., & Maiti, P.C. (1957). *Cordyceps sinensis* (Berkeley) saccardo: Structure of Cordycepic acid. *Journal of The American Pharmaceutical Association, 46*(2), 114–118.

Chen, D.G. (1995). Effects of JinShuiBao capsule on the quality of life of patients with heart failure. *Journal of Administration of Traditional Chinese Medicine*, 5, 40–43.

Chen, G.Z., Chen, G.L., Sun, T., Hsieh, G.C., & Henshall, J.M. (1991). Effects of *Cordyceps sinensis* on murine T lymphocyte subsets. *Chinese Medicine Journal, 104*(1), 4–8.

Chen, Y.J., Shaio, M.S., Lee, S.S., & Wang, S.Y. (1997). Effect of *Cordyceps sinensis* on the proliferation and differentiation of human leukemic U937 cells. *Life Sciences, 60*(25), 2349–2359. doi: 10.1016/s0024–3205(97)00291–9

Chen Marshall, A. (2020). Traditional Chinese medicine and clinical pharmacology. In F. Hock, & M. Gralinski (Eds.), *Drug discovery and evaluation: Methods in clinical pharmacology* (pp. 455–482). Springer: Cham. doi: 10.1007/978-3-319-68864-0_60

Cheung, J.K.H., Li, J., Cheung, A.W.H., Zhu, Y., Zheng, K.Y.Z., Bi, C.W.C., Duan, R., Choi, R.C.Y., Lau, D.T.W., Dong, T.X., Lau, B.W.C., & Tsim, K.W.K. (2009). Cordysinocan, a polysaccharide isolated from cultured Cordyceps, activates immune responses in cultured T-lymphocytes and macrophages: Signaling cascade and induction of cytokines. *Journal of Ethnopharmacology, 124*(1), 61–68.

Congressional Research Service (CRS). (2020). Covid-19: China medical supply chains and broader trade issues. Report R46304. https://sgp.fas.org/crs/row/R46304.pdf

Curran, J. (2008). The Yellow Emperor's Classic of Internal Medicine, *BMJ*, 336, 777.

Hao, T., Li, J., Duu, Z., Duan, C., Wang, Y.M., Wang, C.Y., Song, J.P., Wang, L.J., Li, Y.H., & Wanh, Y. (2012). *Cordyceps sinensis* enhances lymphocyte proliferation and CD markers expression in simulated microgravity environment. *Zhongguo Shi Yan Xue Ye Xue Za Zhi, 20*(5), 1212–1215.

Holliday, J., Cleaver, M., Tajnik, M., Cerecedes, J.M., & Wasser, S.P. (2010). Cordyceps. In P.M. Coates, J.M. Betz, M.R. Blackman, G.M. Cragg, M. Levine, J. Moss, & J.D. White (Eds.), *Encyclopedia of dietary supplement*. Informa Healthcare: NY.

Huang, F., Li, W., Xu, H., Qin, H., & He, Z.G. (2019). Cordycepin kills Mycobacterium tuberculosis through hijacking the bacterial adenosine kinase. *PLOS ONE, 14*(6), e0218449. https://doi.org/10.1371/journal.pone.0218449

Keji, C., & Hao, X. (2003). The integration of traditional Chinese medicine and Western medicine. *European Review, 11*(2), 225–235. https://doi.org/10.1017/S106279870300022X

Khan, M.A., Mousumi, T., Dianzheng, Z., & Chen, H. (2010). Cordyceps mushroom: A potent anticancer nutraceutical. *PCOM Scholarly Papers*. Paper 373.

Koh, J.H., Kim, J.M., Chag, H.J., & Suh, H.J. (2003). Hypocholesterolemic effect of hot-water extract from mycelia of *Cordyceps sinensis*. *Biological and Pharmaceutical Bulletin, 26*(1), 84–87.

Lee, D.Y. (2022). Online presentation at McLean Hospital webinar. Harvard. November 2022.

Liang, H.J., Niu, G.M., & Guan, Y.F. (2006). *Cordyceps sinensis* combined with anti-HBs positive placenta for the treatment of chronic hepatitis B. *Shandong Medicine, 46*(31), 66.

Liang, Z., & Zhao, Z. (2017). The original source of modern research on Chinese medicinal materials: Bencao texts. *Journal of Alternative Complementary and Integrative Medicine, 3*(4). doi: 10.24966/ACIM-7562/100045

Long, H., Qiu, X., Liu, G., Rao, Z., & Han, R. (2021). Toxicological safety evaluation of the cultivated Chinese Cordyceps. *Journal of Ethnopharmacology*, 268, 113600.

National Toxicology Program, Institute of Environmental Health Sciences, National Institutes of Health (NTP) (1992). *National Toxicology Program Chemical Repository Database*. Research Triangle Park, North Carolina.

Paterson, R.M. (2008). Cordyceps: A traditional Chinese medicine and another fungal therapeutic biofactory? *Phytochemistry, 69*(7), 1469–1495. doi: 101016/j.phytochem.2008.01.027

Pierson, D., Qian, I., Wang, O., & May, T. (2022, December 20). China's abrupt Covid pivot leaves many without medicine. *The New York Times*. https://www.nytimes.com/2022/12/20/world/asia/china-covid-shortages.html

Shin, K.H., Lim, S.S., Lee, S., & Lee, Y.S. (2003). Anti-tumour and immuno-stimulating activities of the fruiting bodies of Paecilomyces japonica, a new type of Cordyceps spp. *Phytotherapy Research, 17*(7), 830–833. doi: 10.1002/ptr.1253

Sprecher, M., & Sprinson, D.B. (1963). A reinvestigation of the structure of 'Cordycepic acid' *Journal of Organic Chemistry, 28*(9), 2490–2491. https://pubs.acs.org/doi/10.1021/jo01044a536https://doi.org/10.1021/jo01044a536

Stevenson, A. (2022, November 29). China says it will do more. *New York Times*. https://www.nytimes.com/2022/11/29/business/china-covid-vaccinations.html

Sun, J., An, L., Zhang, Z., Yuan, G., & Du, P. (2017). Extraction methods and sedative-hypnotic effects of polysaccharide and total flavonoids of Cordyceps militaris. *Biotechnology and Biotechnological Equipment*, 2, 495–505.

Turliuc, D., Cucu, A., Costachescu, B., & Tudor, R.M. (2019). The use of mannitol in neurosurgery and neuro-ophthalmology. *Cellulose Chemistry and Technology, 53*(7–8), 625–633.

Unschuld, P.U. (Ed.) (2018). *Traditional Chinese medicine: Heritage and adaptation*. Columbia University Press: NY.

Verma, A.K. (2022). Cordycepin: A bioactive metabolite of Cordyceps militaris and poly-adenylation inhibitor with therapeutic potential against COVID-19. *Journal of biomolecular structure & dynamics, 40*(8), 3745–3752.

Wang, A. (2022). *Zheng Bu Tu ao Bei Yao*. Reprint from 1891 edition. SN Book World: Delhi.

Wang, J., Chen, H., Li, W., & Shan, L. (2021). Cordyceps acid alleviates lung cancer in nude mice. *Journal of Biochemical and Molecular Toxicology*, 35, 22670.

Winkler, D. (2008). The mushrooming fungi market in Tibet exemplified by *Cordyceps sinensis* and Tricholoma matsutake. *JIATS*, 4, 1–46.

Wu, B.Y., Gao, Y.Q., Liu, G.X., & Yuan, S.L. (2007). Toxicity of artificially cultured *Cordyceps sinensis* silk and its anti-fatigue, hypolipidemic effects. *Modern Preventive Medicine, 34*(16), 3096–3097.

Wu, X.V., Chi, Y., Yu, M., & Wang, W. (2020). *Traditional Chinese medicine as a complementary therapy in combat with COVID-19—A review of evidence-based research and clinical practice*. Beijing.

Wu, Y. (2013). Efficacy of Ningxinbao combined with low-dose amiodarone in the treatment of cardiac arrhythmias in the elderly [in Chinese]. *Journal of Practical Chinese Medicine, 29*(7), 567–568. http://www.chinadoi.cn/portal/mr.action?doi=10.3969/j.issn.1004-2814.2013.07.048

Wu, Y. (2013). *Ben Cao Cong Xin* [in Chinese]. China Press of Traditional Chinese Medicine: Beijing.

Xu, C., Wu, D., Zou, Z., Mao, L., & Chunhua, X. (2023). Discovery of the chemical constituents, structural characteristics, and pharmacological functions of Chinese caterpillar fungus. *Open Chemistry, 21*(1). doi: 10.1515/chem-2022–0337

Xu, R.H., Peng, X.E., Chen, G.Z., & Chen, G.L. (1992). Effects of *Cordyceps sinensis* on natural killer activity and colony formation of B16 melanoma. *Chinese Medical Journal, 105*(2), 97–101.

Yang, C.K., Hou, S.Y., & Yu, J. (1990). Treatment of tardive dyskinesia with Ningxinbao capsule. *New Drugs and Clinics, 3*(5), 279.

Yang, S., & Zhang, H. (2016). Production of intracellular selenium-enriched polysaccharides from thin stillage by *Cordyceps sinensis* and its bioactivities, 10, 3402.

Yang, W.Z., Deng, X.A., & Hu, W. (1985). Clinical study on the treatment of hypogonadism by *Cordyceps sinensis*. *Jiangxi Traditional Chinese Medicine*, 5, 46.

Ye, F. (2017). Report from Ouanzhou Hospital. https://m.youlai.cn/yyk/articlemip/312017.html

Yoneyama, T., Takahashi, H., Grudnewska, A., Ban, S., Umeyama, A., & Noji, M. (2022). Ergostane-type sterols from several Cordyceps strains. *Natural Product Communications*, *17*(6), 1934578.

Yu, S.H., Chen, S.Y., Li, W.S., Kumar Dubey, N., Chen, W.H., Chuu, J., Leu, S.J., & Den, W.P. (2015). Hypoglycemic activity through a novel combination of fruiting body and mycelia of Cordyceps militaris in high-fat diet-induced type 2 diabetes mellitus mice. *Journal of Diabetes Research*, 723190.

Yuan, X. (2020). Extraction, structure and pharmacological effects of the polysaccharides from *Cordyceps sinensis*: A review. *Journal of Functional Foods*, 89, 104909. https://doi.org/10.1016/j.jff.2021.104909

Zhang, J.C., Ma, F., & Li, D. (1986). Summary of 30 cases of malignant tumors treated with adjuvant therapy of Zhi Ling capsule. *Shanghai Journal of Traditional Chinese Medicine*, 10, 25. doi: 10.16305/j.1007-1334.1986.10.015

Zhang, J., Yu, Y., Zhang, Z., et al. (2011). Effect of polysaccharide from cultured Cordyceps on immune function and anti-oxidation activity of mice exposed to 60Co. *Silva Gandavensis*, *11*(12), 2251–2257.

Zhang, S.L. (1920). *Ben Cao Zheng Yi* [in Chinese]. Shanxi Science and Technology Publishing House: Taiyuan.

Zhang, Y. (1871). *Ben Cao Mu Shi Yi* [online reprint]. Library of Congress. Chinese Rare Books Collection. https://www.loc.gov/resource/lcnclscd.2012402163.1A000/?st=gallery

Zhao, M., Wang, K., Gu, R., & Zhong, S. (2019). A comparative study on shared-use medicines in Tibetan and Chinese medicine. *Journal of Ethnobiology and Ethnomedicine*, 15, 43.

Zhu, F. (1767). *Gan Yuan Xiao Shi Ben Cao Cong Xin* [in Chinese].

Zhu, J.S., Halpern, G.M., & Jones, K. (1998). The scientific rediscovery of an ancient Chinese herbal medicine: *Cordyceps sinensis*: Part I. *Journal of Alternative & Complementary Medicines*, 4, 289–303.

Zhu, X.Y., & Yu, H.Y. (1990). Immunosuppressive effect of cultured *Cordyceps sinensis* on cellular immune response. *Zhong Xi Yi Jie He Za Zhi*, *10*(8), 485.

Zhuan, X.K. (1969). *Liu ya wai bian*. Chu ban. Taibei Shi: Guang wen shu ju, Minguo, 58.

Part II

Supply Chains of *Cordyceps sinensis*

A firm should create value for its buyers that exceeds the costs spent, in order to sustain competition in the market and create profit. Successful firms make successful industries. The traditional approach to industrial analysis is to focus on the industrial structure and competitors. Porter (1985)[1] created a framework (Porter's Diamond) that highlighted potential risks that should be addressed by a firm in strategic planning: rivalry among existing firms, potential entrants, bargaining power of suppliers and buyers, and potential substitutes.

From this perspective, the *Cordyceps sinensis* industry now stands at a crossroads: a potentially cheaper substitute is *Cordyceps militaris*, as it is grown artificially, and with the arrival of cheaper online communication and quick delivery, existing suppliers can bypass intermediary agents and sell directly to consumers. But there is an even more important threat—the natural supply of *Cordyceps sinensis* is rapidly diminishing because of climate change and overharvesting. On the surface, this is bad news for industrial players. However, as the artificial production of *Cordyceps sinensis* has been successful, it is only a matter of time before key players in the industry turn to artificial production. However, creating value is not only about the competitiveness of one firm or a whole industry, or simply about focusing on the benefits to the customer from consuming the products. Product creation has much wider socio-economic and environmental implications, which can be identified in analyzing value at each stage of production. From this perspective, a sustainable value chain analysis is useful. Cordyceps is a medicinal food product, so The Food and Agriculture Organization of the United Nations's definition of a sustainable food value chain is the most applicable. FAO defines a sustainable value chain as the 'successive and coordinated value-adding activities' of several actors, including farmers and firms. These value-adding activities should be 'profitable throughout all of its stages' (economic aspect of sustainability), create 'broad benefits for society' (social aspect sustainability), and should not 'permanently deplete natural resources' (ecological aspect of sustainability) (Neven, 2014).[2]

In the *Cordyceps sinensis* industry, the development of a value chain is very much about managing the supply side. Here lie the most significant bottlenecks—a diminishing natural pool of caterpillar fungi and too much dependence on the income primary collectors make from harvesting it. The complete transition to artificial Cordyceps in the industry will push thousands of families into deep poverty

DOI: 10.4324/9781003429753-5

in four Himalayan countries. Intermediary agents are also vulnerable to the recent changes, but some of them have applied proactive strategies to address new trends in the market. Thus, in this part of the book we have focused on analyzing supply chains of *Cordyceps sinensis*. Chapter 3 explains the geography and processes of harvesting wild Cordyceps in China. Chapter 4 will provide details on the organization of trade in Cordyceps between primary collectors and trading agents. Chapter 5 provides a review of the existing Cordyceps-based products in the Chinese market and lists key production companies and analyzes consumer profiles. Finally, Chapter 6 explores placing the Chinese *Cordyceps sinensis* industry in the global dimension and discusses potential competitors—suppliers from Bhutan, India, and Nepal and opportunities for entering new markets (Russia).

Notes

1 Porter, M. (1985). *Competitive advantage: Creating and sustaining superior performance*. The Free Press: N.Y.
2 Neven, D. (2014). *Sustainable food value chains: Guiding principles*. FAO: Rome.

3 The Where, Who, and How in Collecting Wild Cordyceps in China

Jiping Sheng, Xinyi Xu, Yiyuan Miao, Lin Shen, and Ksenia Gerasimova

Geographically, Cordyceps are harvested across the Qinghai-Tibetan Plateau, but the ethnic background of Cordyceps collectors varies locally, and the boundaries between them do not match the official administrative units. Below we provide details of the key locations for harvesting wild Cordyceps in China and approximate harvested outputs, and compare the socio-economic status of those who collect wild Cordyceps in these areas, addressing local differences in harvesting the precious fungi.

Geographic Distribution of Collecting Cordyceps

Gansu Province

Cordyceps sinensis in Gansu Province is mainly distributed near the Qinghai Lake, Dunhuang, Jiayuguan, Zhangye, Gannan Longnan, Dingxi, Lanzhou, Tianshui, Qiyang, Pingliang, Wuwei, and Zhangye. The Gansu annual wild Cordyceps production capacity is 8.[1] tons. The province is populated by diverse ethnic groups. Gannan Tibetan Autonomous Prefecture is the main production area in Gansu, accounting for more than 90% of the output of *Cordyceps sinensis* in Gansu Province (Qinghai Cordyceps Network, 2011).

The Gannan Tibetan Autonomous Region is located south-west of Gansu. The Tibetans of the Gannan consist of two different groups. Diebu (Tebo, Tewo) animal herders live in the northern part of the territory, amid vast grasslands, while Zhouqu live in the south at lower altitudes and engage in planting crops as the main source of subsistence (Kang et al., 2016).

The Qilian Mountains, known as Nan Shan, rise to 19,000 feet (5,800 meters), with most peaks exceeding 14,000 meters. Tianshan Mountain is the ancient Chinese name for the Qilian Mountains. Its glaciers occupy a territory of almost 2,000 square km. It forms a natural border between two provinces—Gansu and Qinghai—and the Hexi Corridor in the impassable ragged mountains was a famous passage serving as a border between civilizations, separating Qiang, Hall, Tubo, who are ethnically close to the Tibetans, in the south, and the Hui, the Ugurs, and Mongolians in the north. The ethnic and cultural differences, distinguishable in appearances, language, and music, are now blurred; the only visible difference

DOI: 10.4324/9781003429753-6

is the traditional lifestyle. In the eyes of local nomads, ethnic background is less important than the lifestyle: 'The steppe is not divided into mountains and corridors. The only thing that exists is pasture, just the huge "muddy land" (Hundi, Changchuan) of the Qilian Mountains and the north of the mountain' (Mongolian Herder, cited in Chong, 2020).

Heihe River Basin (HRB), known as the 'gift of the Qilian Mountain,' is the second largest inland river basin in China and contains 17 ecological sites and 35 key ecological corridors. The largest ones are the Qilian Mountain Nature Reserve and the Heihe National Wetland Park (Wu et al., 2023). The area is also a hotspot for biodiversity.

The intense extraction of resources was practiced in the region from the 1960s until 2017, when the Qilian Mountain National Reserve was created as a pilot project to identify the best practices in environmental protection and to stop harmful practices (Li et al., 2022).

The reserve is located in Sunan County, Gansu Province, which is the only Yugur Autonomous County in the country, with the capital city Zhangye. It is located in the middle of the Hexi Corridor, at the northern foot of the Qilian Mountains. It belongs to the Heihe Ecological Restoration and Conservation Development Zone. It has a cold and semi-arid climate, with a long winter season and short summer. The local population still relies on local wild plants and mushrooms for food and medicinal treatment; the older generation has a better knowledge of varieties and usage (Kang et al., 2016). Medicinal valuable plants, apart from Cordyceps, include Tibetan rhubarb, Notopterygii Rhizoma et Radix (Qian-huo), snow lotus, Radix Codonopsis, and Radix Bupleuri.

Due to its ecological and industrial importance, the Sunan local government introduced regulations on harvesting local Cordyceps. Collecting Cordyceps is an additional income which requires high investment in time and stamina. Beginning in May each year, for a period of about one month, local farmers and herdsmen go to the grassland to collect Cordyceps, often just by themselves. Cordyceps collectors use an L-shaped digging tool with a length of about 10 cm at both ends, a width of about 2 cm, and a thickness of about 5 mm, which is convenient to carry.

> Fungi can be found underground, at a very shallow level, and people who dig Cordyceps usually use a slender small gardening pickaxe (about 20 cm long). It is light and does little damage to the grassland. When digging, the person uses the tip of the pickaxe to lift the turf near the side of the *Cordyceps sinensis*, lifts the fungi gently, and puts the soil clod into the original place and gently compacts it. Cordyceps that do not look good should not be dug, because it has no medicinal value and should be reserved for the coming year.
>
> (Interview with the Sunan farmer, Xining, May 2018)

As the regulations for harvesting wild plants and fungi in the environmentally protected areas are strict, only local residents are allowed to collect Cordyceps. To get a better price, individual Cordyceps collectors from Sunan may sometimes go to Xining, a commercial hub for raw Cordyceps sales (Interview with Tibetan herder, Xining, May 2018). For a long time, Sunan rural households consisted of three types: households involved only in agricultural activities, those in non-agricultural activities, and those in mixed activities. Under climate change and deteriorating pastures, local herders are experiencing significant pressures. Those who are living within the reserve and rely on herding are found to be the most vulnerable groups, despite getting certain conservation-related jobs inside the reserve, while those who come from the peripheral areas of the reserve have the highest adaptability rates and have outside jobs. Collecting plants and fungi is still an important survival strategy for rural Sunan communities (Li et al., 2022).

Qinghai Province

Qinghai Province in its current form was organized in 1928. Historically, it was part of Amdo, the Tibetan kingdom. This is still reflected in the ethnic background of its population. The province is a melting pot, where different ethnicities, such as Tibetan pastoralists and farming Hui and Han live together. Xining and Handong Districts are the most urbanized and have the highest concentration of Han communities. The rest of Qinghai is predominantly Tibetan. According to the 2000 census, Tibetans account for 66 percent of the population of Malho TAP, 63 percent of Tsolho TAP, 92 percent of Golog TAP, and 97 percent of Yushu. Other minorities in the region: Mongolian, Mongour, and Qiang also speak the Tibetan language (Fischer, 2008).

The Qinghai areas which are predominantly inhabited by the Tibetans include grasslands that produce Cordyceps. Qinghai Province has the largest number of pastures that produce wild Cordyceps in China, and of high quality. In the early 1980s, Qinghai's output was 30 tons. Only Yushu and Golog's output accounted for more than 85% of the total output of Qinghai Province, with Yushu Prefecture's output ranking first (Qinghai Spring Medicinal Resources, 2016). In addition, Cordyceps is also harvested in Xinghai County, Tongde County, Guide County of Hainan Tibetan Autonomous Prefecture, Tongren County and Henan County of Huangnan Tibetan Autonomous Prefecture, and the mountainous areas in Hulong County, Huzhu County and Minhe County of Haidong District, but the output is little. Among them, the quality of Yushu Cordyceps is ranked the best, followed by the Golog Cordyceps (Interview with Cordyceps trader, Xining, May 2018).

Somehow there is a competition between those who are from Qinghai and TAR for the reputation of having the best local variety of Cordyceps. A Qinghai Cordyceps trader complained:

> Each year, *Cordyceps sinensis* is produced: 70% in Qinghai and only 20% in Tibet. But the propaganda of the Tibetan government is very good; Tibet has a unique sense of mystery, ethnic customs, and geographical location, so everyone is full of mystery and yearning for Tibet, so people subconsciously think that the Cordyceps of Tibet is so good, but in fact, what

> many Tibetan locals eat is Qinghai Yushu Cordyceps, the best grass worm that can be found.
>
> (Interview with Qinghai Farmer, Xining, May 2018)

On average, snow in the Qinghai-Tibet Plateau begins to melt between May and June each year, which is considered the best time for harvesting Cordyceps in Qinghai Province:

> At this time, the fungi stromata are exposed to the snow surface while the seedlings do not exceed one inch. At the best digging time, not only the worm body is full and fat but the grass head is short, so it is easy to find and dig. The amount, quality, and efficacy are important. If you dig too early, most of the Cordyceps are still underground, and some are even worms, which are difficult to find and dig. If you dig too late, the snow will melt, and the weeds will grow. The Cordyceps will not be easy to find, and the seedlings will die, the worm body inside the soil will wither and will not be suitable for medicinal use. The production area of wild Cordyceps has a digging period of about 12–26 days between May and June. In areas with relatively low altitudes, if the digging period starts earlier, the quality of Cordyceps worsens.
>
> (Interview with Qinghai Farmer, Xining, May 2018)

Collecting Cordyceps among rural residents of Qinghai has become popular since the late 1980s. Up to the late 1990s, the total output of Cordyceps in China was 100 tons a year, with Qinghai contributing 50% of all harvested Cordyceps. In the 2000s, the total production output increased to 2,000 tons a year, and Qinghai Cordyceps accounted for 60% (Qinghai Spring Medicinal Resources, 2016). Harvesting Cordyceps has also become the second (after animal husbandry) largest source of income for farmers and herdsmen in Qinghai. For more than 20 years of 'crazy digging' for Cordyceps, also known as 'gold mushrooms,' farmers have been caught in a vicious circle of 'the more you dig, the less the amount; the less the amount, the high the price; the high the price, the more you dig' (Interview with Qinghai Farmer, Xining, May 2018).

The Qinghai cordyceps collectors also organized the Qinghai Provincial Cordyceps Association, which was registered and established at the Qinghai Provincial Department of Civil Affairs on September 19, 2007. It is a non-profit group of Cordyceps collectors, individual (private) operators, and brokers in the province, 'privately run, privately managed, and privately benefited' and operating on the principles of 'self-education, self-management, and self-service.' Its purpose is to link collectors with industry—i.e., enterprises that process raw Cordyceps. According to document No. 36 Guobanfa (2007) of the General Office of the State Council, 'Several Opinions on Accelerating the Reform and Development of Industry Associations and Chambers of Commerce' and the Qinghai Provincial Government General Office Qingbanfa (2007) No. 178 'On the Reform and Development of Industry Associations and Chambers of Commerce,' the association stipulates voluntary regulation of the Cordyceps industry and coordination to promote research and development and protect Cordyceps products as Qinghai regional products (Qinghai Provincial Cordyceps Association, 2023).[1]

Tibetan Autonomous Region

In September 1965, the First Session of the First People's Congress of the Tibet Autonomous Region proclaimed the establishment of the Tibet Autonomous Region. TAR has the largest Tibetan population in China at 2.7 million people, followed by the Sichuan and Qinghai Provinces. Despite calls for a standardized version of Tibetan, there are still three main languages being practiced in TAR: Dbus-gtsang, Kham, and Amdo (Hartley, 1996). The areas that produce the most wild Cordyceps are distributed throughout the territory of the Tibet Autonomous Region, accounting for more than 56% of its total area.[2]

By the end of April each year, the Cordyceps grow consistently. When the temperature rises to about 10°C in May, the stroma begins to drill out of the ground, and after around 10 days of growth, it can be harvested (Interview with Tibetan farmer, the Namso Lake, May 2018). Harvesting Cordyceps has become one of the main sources of income for rural communities in many counties, with an annual total income value of more than 10 million yuan. The annual transaction volume of Cordyceps is about 8 tons, the annual output value is more than 55 million yuan, and the income generated accounts for more than 60% of the country's GDP (Fanhonghua, 2020).

Local farmers and herdsmen are mostly Tibetans, and their right to harvest local Cordyceps is protected by national and local legislation. The grasslands where the Cordyceps are produced are mostly distributed among the Tibetan herders' households. The mountains or grasslands that are not allocated to households are often owned by the village collectives, who control the harvesting of Cordyceps. Therefore, only local herders or village collective members can lawfully collect Cordyceps. Generally, before digging in the pasture, the potential Cordyceps collector should obtain 'the digging certificate' issued by the government, and the certificate should also be qualified by the residents or their relatives.

Sichuan

Sichuan Province is producing 20% of Chinese wild Cordyceps. Its Western region, which is on the edge of the Qinghai-Tibet Plateau, consists of mountainous areas with an altitude between 3,000 and 5,000 meters. Almost in all counties in Ganzi Tibetan Autonomous Prefecture, Aba Tibetan Autonomous Prefecture, and Qiang Autonomous Prefecture, the local Tibetan population harvests wild Cordyceps (Tibet Nagqu Cordyceps Network, 2022).

The Ganzi (Kardze Prefecture) and Aba varieties of *Cordyceps sinensis* are the most famous. Geographical conditions in Kardze Prefecture are similar to those in the Qinghai-Tibet Plateau, characterized by low temperature, long winters, little precipitation, and sufficient sunshine, which is suitable for the growth of Cordyceps. Aba Prefecture is on the southeast edge of the Qinghai-Tibet Plateau, adjacent to Yushu, Qinghai, forming the hinterland of the Qinghai-Tibet Plateau, thus both produce wild *Cordyceps sinensis* of comparable quality (ibid.).

Sichuan Cordyceps is different to Cordyceps produced in Tibet and Qinghai. The Sichuan variety has 'red or dark brown eyes' and a lighter taste. It contains the same active ingredients, but the nutritional value is slightly lower, which is reflected in the price. Cordyceps are harvested specifically in Kangding, Sêrxü County, and Baiyu County (Li & Cao, 1990).

Sichuan Cordyceps-producing areas also host famous monasteries such as the Chonggu Monastery, the most visited Buddhist monastery in the county, in the Yading Nature Reserve. Despite formal (nature reserve regulations) and informal (monks' guardianship, religious taboos) nature protection regimes, the area still sees an influx of people, including pilgrims, collecting Cordyceps and disturbing rare species such as the white eared pheasant (Crossoptilon crossoptilon) (Wang et al., 2012).

Yunnan

The southernmost area for harvesting wild Cordyceps in China is Yulang Naxi County in Yunnan Province. The area known as 'the three parallel rivers': tributaries of the Yangtze river, Jinsha river, Lancang and Nu rivers consists of 8 clusters of 15 protected areas which form an epicenter of Chinese biodiversity (UNESCO, 2010). Baima Xueshan Nature Reserve was established in 1983 mainly to protect populations of the Yunnan snub-nosed monkey (Rhinopithecus bieti), the only primate that can live at high altitudes and under low temperatures. It is prohibited to harvest plants and Cordyceps in the reserve, but since the local population, mostly of Tibetan origin, does not have alternative income generation activities, the nature reserve authorities allow limited harvesting of Cordyceps, but implement strict management measures (Weckerle et al., 2010).

The diverse landscape includes the Biluo Mountains (Lanping), Laojunshan Mountains, Meili Mountains, Haba Mountains, and Gongshan Mountains, which provide suitable high-altitude habitats for Cordyceps (Dai et al., 2020). In Yunnan, wild Cordyceps are harvested in Diging, Nujiang, and Lijiang, where five new species were identified. They are morphologically distinct from all other Cordyceps sensu lato, so it was suggested that they should differ from the others in the genus, based on combined multigene analyses (Dong et al., 2022).

Comparing Local Variations in Harvesting Cordyceps

Harvesting Calendar and Equipment

After investigating different production areas, we found that the practice of collecting Cordyceps varies across the different regions (Table 3.1).

The harvesting period of wild Cordyceps is short, usually from early May to mid-June. After mid-to-late June, Cordyceps enters the spore emission stage, and the mountain protection regulation prohibits collecting such Cordyceps. Removal

Table 3.1 Characteristics of local practices of Cordyceps collection

	Sichuan	*Gansu, Qinghai, and Xizang*
Time	Starting from April to May of each year	Gansu: Starting from May of each year and lasting for one month Qinghai and Xizang: Starting every April and continuing until June
Number of collectors	In groups or alone	In small groups or alone
Careful removal of each Cordyceps?	Yes	Yes
Digging Tools	Traditional appliance with a length of about 20 cm and a width of about 9 cm, equipped with wooden handle.	L-shape appliance with a length of about 10 cm, a width of about 2 cm and a thickness of about 5 mm; it is convenient to carry.
Details of digging process	The same digging location can be continuously excavated for several years, and gradually deep digging every year forms a deep pit of 1–2 meters.	The tools are small and the farmers and herdsmen leave almost no visible traces of digging.

of the complex fungus body is a delicate matter, and it requires special equipment in order not to destroy the mummified body of the worm. Collectors usually use a small slender gardening pickaxe (no more than 20 cm long), and it is easy to remove just a small clod of soil. This tool causes very little damage to the grassland. Our interviewees mentioned several times the requirement to put back 'the turf' after picking up the stroma of Cordyceps. They also know that Cordyceps that has been atrophied and emptied (commonly known as 'chemical seedlings') should not be collected, because it has lost medicinal and commercial value.

During our fieldwork, we did not come across any cases where machinery was used to harvest Cordyceps—it was all meticulously hand-picked—but the idea and patented designs of robotics and machinery that can partially replace people in the harvesting of Cordyceps in the Tibetan meadows have been circulating for the past 2–3 years. Firms such as Qufu Huilin Machinery from Jining, Shandong, have marketed their small (170 x 70 x 85 cm) soil excavator normally used for digging potatoes also as a Cordyceps harvester (Qufu Huilin Machinery, 2023).[3] The price for such machinery was 2,580 RMB in July 2023. Another company, Zhuoyue Zonglian in Chengdu, Sichuan, patented machinery for searching for distant Cordyceps in the grasslands. The automatic system includes a camera for collecting ground images, a mobile robot with a driving device, and a computer chip with installed image detection and recognition technology to avoid obstacles in the ground and control the robot's position. It also has a GNSS chip to recognize the global geographical location, an indicator light, a data chip, and a power

Figure 3.1 Farmers and herdsmen digging Cordyceps

supply module (Zhuoyue Zonglian Company, 2023).[4] The company claims that its technology can promote automated agricultural production, reduce labor costs, improve the efficiency of the fungi search process, and protect the vegetation of grassland meadows.

Different Ways of Contracting Cordyceps Collection

Collecting Cordyceps requires certain stamina, familiarization with the local landscape, and much patience, as collectors search at high altitudes for hours and days. There are also health and safety risks, so hunting for Cordyceps requires preparation.

Two social practices in collecting wild Cordyceps have been identified: the individual hunt for Cordyceps or collective (group) harvesting. Such practices are determined by the land rights, resource abundance, and the pasture access regime. At the peak of the Cordyceps gold rush, an uncontrolled flow of people came to the highland pastures and contributed to the overharvesting of Cordyceps. To differentiate outsiders from the locals, who still needed the extra income, and control how much is harvested, the two regions with the highest Cordyceps output (Qinghai Province and TAR) have introduced similar schemes to regulate access to Cordyceps collection. 'The Interim Measures for the Collection and Management of Cordyceps' were introduced on January 1, 2005, and April 1, 2006, in Qinghai Province and TAR respectively. In 2023, Qinghai Province updated its policy on collection and sale.

For better environmental protection and traceability of raw materials, the forestry and grassland bureaus of each county will, in conjunction with the township governments, scientifically formulate the annual plan for harvesting Cordyceps resources according to meteorological conditions and vegetation conditions, and reasonably determine the area, the suitable excavation amount, the number of digging personnel, the digging period and relevant safeguard measures to ensure the rational development and utilization of Cordyceps resources. Since 2023, units or individuals that start selling *Cordyceps sinensis* need to bring to the forestry and grassland department of the county (city, district, and administrative committee) the application form for administrative licensing of wild plants under national key protection, a copy of the collection certificate, a copy of the purchase contract, and a copy of the sale contract. They will issue a preliminary review opinion and report to the forestry and grassland bureau of Qinghai Province for approval.

Both regions' authorities introduced a certification scheme for residents intending to collect Cordyceps. According to the regulation, local people within the county area can officially apply to their local township people's government to collect Cordyceps. The agricultural and animal husbandry administrative departments of counties (cities, districts) entrust these township people's governments to issue Cordyceps collection certificates. Usually, this scheme is implemented for grasslands which are listed under collective land ownership rights.

At present, there is no provincial-level certification in Sichuan, and Yunnan, but there are smaller, corresponding management systems at the regional level, such as 'The Interim Measures for the Collection and Management of Cordyceps in Ganzi Tibetan Autonomous Prefecture, Sichuan.' In May 2023, Gansu Province introduced stricter controls that made local authorities and farmers responsible for grassland guardianship (Interview, Cooperative Municipal Bureau of Natural Resources, 2023).

According to rough estimates of Cordyceps dealers in the Xining market, around 2010 approximately 80%–90% of the total output of Qinghai-Tibetan Cordyceps was collected by herdsmen, and revenues received from Cordyceps collection and trade accounted for 60%–80% of the total income of herders. The turning point seemed to be the year 2014, when the output of Cordyceps in Qinghai Province was about 35–40 tons; it was only about 40% of that in 2013. The outputs of Yushu, Golog, Hainan, and Huangnan counties in the main production areas have been declining; but the outputs in scattered production areas such as Haibei, Haidong, and Xining have increased, probably due to less overharvesting than in the traditional Cordyceps rich areas (Interview with Cordyceps dealer, Xining, May 2018). In the best years, according to Cordyceps dealers, a person could collect hundreds of Cordyceps in a day from the Yushu Tibetan Autonomous Prefecture, an important production area in Qinghai. As a result, at the end of the harvesting season, the whole family could have earned ten thousand yuan or even hundreds of thousands of yuan. However, in the following years, people who could dig 10 Cordyceps 'sticks' in one day were considered lucky (Interview with Yushu herder, Xining, May, 2018).

Under the land rights regime, individual households own rights to the pastures where Cordyceps are collected. They have a choice: they can either harvest Cordyceps themselves, hire workers to do the harvesting, or even rent out the pasture to another 'boss,' commonly known as a Cordyceps mountain contractor (Interview with Cordyceps dealer, Xining, May 2018). In Qinghai Province, Cordyceps pastures are commonly owned by Tibetans, who rent them out to Hui mountain contractors, or a middleman, who invests in the Cordyceps harvesting, including mountain rent (at least 200,000–300,000 yuan), and employs seasonal workers, who receive food subsistence and are paid per each *Cordyceps sinensis* piece collected (6–20 yuan) (ibid.).

Cordyceps mountain contractors are generally Hui people, while pastures are generally owned by the Tibetans. This model of pasture management is widespread in Qinghai but not in the other regions, such as Gansu and TAR.

In Tibetan counties in Qinghai and TAR, it is very common for schools to schedule special 'Cordyceps collecting' holidays, which involve releasing students from elementary and high schools so that they can assist their families in the collective hunt for Cordyceps.

Larger sites for collecting wild Cordyceps, such as Agagou, in Litang, Sichuan Prefecture, one of the largest sites in the county, can simultaneously host around 10,000 people, who make up a temporary community. They are driven by trade worth millions of USD and the developed infrastructure ('the camp roads are cement-hardened, and two roads have been built up to the mountains at an altitude of 5,000 meters') (Red Star News, 2022). Such a large group of people is prone to many conflicts, and this creates potential hazards. Local police have to organize temporary commissaries to be stationed at the Cordyceps collecting sites to stop gambling, alcohol consumption, and fighting:

> Drinking alcohol can trigger altitude sickness and can also trigger public order cases. The problem of playing mahjong is even more serious, and some young people may lose the Cordyceps that they have worked hard to dig for more than a month.
>
> (Wu Jinqu, cited in Red Star news 2022)

Yushu Cordyceps collection sites are also facing an increased number of conflicts, so the local authorities delegated the Xiaosumang Township Police Station of the Municipal Bureau to organize a crime prevention campaign by informing Cordyceps collectors and running on-site raids (Yushu City Public Security Bureau, 2022)

Due to the uneven distribution of Cordyceps resources and complicated schemes of access to pastures, local disputes between farmers or between villages are rather common. If a village has Cordyceps of better quality, they block access to their pastures, even if the pastures are not technically under their ownership but part of a nature reserve (Weckerle et al., 2010).

Apart from impersonal conflicts, Cordyceps collectors also encounter natural environmental risks. Cases of natural disasters occurring while collecting

Cordyceps are common, and local police forces have stepped in to warn local collectors about potential dangers. For example, climbing in the mountains and holding metal equipment in a thunderstorm is dangerous. In May 2020, in Shiqu county, Kardze, Sichuan, two families were killed by lightning (Xiaoxiang Morning News, 2022). In May 2023, an avalanche injured and trapped Cordyceps collectors near Yangqiao in Guodong Village (China Police Network, 2023). Local people used to attribute the sudden deaths of Cordyceps collectors to supernatural forces punishing human greed with karma (Sulek, 2016), but local police cite the increased frequency of natural disasters, probably caused by climate change and Cordyceps collectors not reading weather predictions and health and safety precautions.

Cleaning Fresh Cordyceps

A distinct gender division in harvesting and processing raw Cordyceps has been recorded. In Tibet, it is mostly males who harvest Cordyceps (He, 2018), or are at least part of a family search. Similarly, in Yunnan 2 in 3 Cordyceps collectors are male (Weckerle et al., 2010). Cleaning the freshly collected Cordyceps is a job for women. Women, mostly Hui and Tibetans, carefully wash and brush the Cordyceps by hand, removing fibrous attachments and impurities. After cleansing, intact Cordyceps are air dried, losing 60–70% of their weight. Once fully dried, the Cordyceps are sprinkled with yellow wine to make them soft, and then tied in a bundle of 6–8 strips with red rope. Then they are tied again in a larger bundle of 200–300 g weight and dried again by a charcoal fire. Women do these procedures at home, in the evening after chores, or at the market, where they are sold.

From our observations, the traditional processing method for preparing fresh Cordyceps for sale consists of four steps:

Step 1: Careful cleaning with a brush to remove the remaining mud. Once mud has dried, it is difficult to remove and might lower the value of the piece.
Step 2: Cordyceps are left to dry in the open air but in the shade. Pieces need to dry almost completely (90%).
Step 3: Then comes the process of sorting the Cordyceps according to quality. Usually, there are six classes: king of kings, king, first class, second class, and 'the rubbish' (usually broken ones). The most expensive can be sold for sixty or seventy yuan or even hundreds of yuan per piece, and the cheapest ones for less than ten yuan.
Step 4: Final checks are given to identify any remaining dirty pieces before sale.

The first two steps are usually completed by the Cordyceps collector. The latter two tasks are usually contracted by the Cordyceps wholesaler or retailer. We have seen many wives of Cordyceps traders or female herders working in the market to sort out and clean cordyceps.

Cordyceps collectors have found that they can add extra weight to a bundle of Cordyceps if they insert twigs or wires into tell-tale holes in the mummified worm body. It is not harmful, but obviously does not contain any medicinal value.

To conclude, most wild Cordyceps collectors are rural residents, often of an ethnic minority background (majority are Tibetans), and rely on harvesting Cordyceps to make a living. In areas such as Gansu and Yunnan provinces, Cordyceps grow in the environmentally protected areas, where de jure harvesting of wild plants and fungi is prohibited, but in practice local residents are allowed to harvest sustainably, as they don't have any viable alternatives to selling Cordyceps, but even sustainable harvesting is damaging the nature reserves. In key production areas such as Qinghai and Tibet, overharvesting is very visible, and local Cordyceps collectors despair over collecting 'the golden fungus' with the hazards of climbing high altitudes, especially during the storm season.

Notes

1. See https://baike.baidu.com/item/青海省冬虫夏草协会/61831133?fr=ge_ala.
2. The distribution areas include Linzhou County, Daban County, Nimu County, Mozhugongka County, Dulong Deqing County, Nagqu County, Jiali County, Baqing County, Nie Rong County, and Su County; Chengdu County, Gongjue County, Basu County, Bianba County, Luolong County, Wuqi County, Chaya County, and Naidong County, Qiongjie County and Cuomei County of Shannan District, Jiacha County, Gongga County, Sangri County, Zhacao County, Longzi County, Langkazi County, Linzhi County, Lang County, Milin County, Bomi County, Gongbu Jiangda County, Xigaze City, Jilong County, Xietongmen County, Nanlinmu County, Bailang County, Jiangyan County, and Yadong County (Tibet Nagqu Cordyceps Network, 2022).
3. See https://www.shihuantong.com/mj/76807/.
4. See httpes://www.qcc.com/firm/9235137b0606136c1080948e6a7071d3.html.

References

China Police Network (2023, 29 May). People digging Cordyceps suddenly caught by an avalanche [in Chinese]. *China Police Network Baidu Official Account*. https://weibo.com/2568309141/N2NnFk4Go

Chong, L. (2009). Zhang Chengzhi: The song of the Xiongnu—The nomadic civilization of the Qilian Mountains and the rise and fall of the Hexi Corridor. *Green Leaf*, 2, 1–2.

Dai, Y., Wu, C., Wang, Y., Huang, L., Chen, Z., Zeng, W., Wang, Y., Yang, Z., Lemetti, P., Mo X., & Yu, H. (2020). Evolutionary biogeography on Ophio*Cordyceps sinensis*: An indicator of molecular phylogeny to geochronological and ecological exchanges. *Geoscience Frontiers*, *11*(3), 807–820. https://doi.org/10.1016/j.gsf.2019.09.001

Dong, Q.Y., Wang, Y., Wang, Z.Q., Tang, D.X., Zao, Z., Wu, H.J., & Yu, H. (2022, 13 April). Morphology and phylogeny reveal five novel species in the genus Cordyceps (Cordycipitaceae, Hypocreales) from Yunnan, China. *Frontiers in Microbiology*. https://doi.org/10.3389/fmicb.2022.846909

Fanhonghua (2020, 15 October). Distribution of *Cordyceps sinensis* resources in Tibet Autonomous Region [in Chinese]. *Fanhonghua*. https://www.fanhonghua.net/chongcao/6071.html

Fischer, A.M. (2008). 'Population invasion' versus urban exclusion in the Tibetan areas of Western China. *Population and Development Review*, *34*(4), 631–662.

Hartley, L.R. (1996). The role of regional factors in the standardization of spoken Tibetan. *The Tibet Journal*, *21*(4), 30–57.

He, J. (2018). Harvest and trade of caterpillar mushroom (Ophio*Cordyceps sinensis*) and the implications for sustainable use in the Tibet region of Southwest China. *Journal of Ethnopharmacology*, 221, 86–90.

Kang, J., Kang, Y., Ji, X. et al. (2016). Wild food plants and fungi used in the mycophilous Tibetan community of Zhagana (Tewo County, Gansu, China). *Journal of Ethnobiology and Ethnomedicine*, 12, 21. https://doi.org/10.1186/s13002-016-0094-y

Li, J., Ma, G., Feng, J., Guo, L., & Huang, Y. (2022). Local residents' social-ecological adaptability of the Qilian Mountain National Park Pilot, Northwestern China. *Land*, 11, 742. https://doi.org/10.3390/land11050742

Li, Q.S., & Cao, Z.Q. (1990). Investigation on ecology of *Cordyceps sinensis* [in Chinese]. *Journal of Chinese Medicine & Materia*, 13, 3–7.

Qinghai Cordyceps Network (2011). Gansu Province Cordyceps resources survey project successfully completed. https://www.123chongcao.com/Html/chongcaozhuanti/86558793101131.html

Qinghai Spring Medicinal Resources (2016). Distribution of *Cordyceps sinensis* in Qinghai [in Chinese].

Red Star News (2022, 20 June). Investigation of the Cordyceps 'gold diggers' on the mountain at an altitude of 4500 meters. *Chengdu Business Daily Red Star Official Account*. https://baijiahao.baidu.com/s?id=1736125418699870312&wfr=spider&for=pc

Sulek, E.R. (2016). Caterpillar fungus and the economy of sinning: On entangled relations between religious and economic in a Tibetan pastoral region of Golog, Qinghai, China. *Etudesmongolese et siberiennes, centrasiatique et tibetaines*, 47, 1–17. https://doi.org/10.4000/emscat.2769

Tibet Naqu Cordyceps Network (2022). Which region of Tibet is Cordyceps best? https://www.xznqcc.com/zixun/2022-08-09/780.html

UNESCO (2010). Three parallel rivers of Yunnan protected areas. https://whc.unesco.org/en/list/1083/

Wang, N., Zheng, G., & Mcgowan, P.J.M. (2012). Pheasants in sacred and other forests in western Sichuan: Their cultural conservation. *Chinese Birds*, *3*(1), 33–46. doi: 10.5122/cbirds.2012.0003

Weckerle, C.S., Yang, Y., Huber, F.K., & Li, Q. (2010). People, money, and protected areas: The collection of the caterpillar mushroom Ophio*Cordyceps sinensis* in the Baima Xueshan Nature Reserve, Southwest China. *Biodiversity Conservation*, 19, 2685–2698. doi: 10.1007/s10531-010-9867-0

Wu, Y., Han, Z., Meng, J., & Zhu, L. (2023). Circuit theory-based ecological security pattern could promote ecological protection in the Heihe River Basin of China. *Environ Science & Pollution Research International*, *30*(10), 27340–27356. doi: 10.1007/s11356-022-24005-5

Xiaoxiang Morning News (2022). Seven people in Sichuan who went up the mountain to dig for Cordyceps were killed by lightning, local: High altitude, extreme weather is common [in Chinese]. https://baijiahao.baidu.com/s?id=1734222341845004454&wfr=spider&for=pc

Yushu City Public Security Bureau (2022). Cordyceps 'season' Yushu City Public Security Work Record (10) Xiaosumang Township Police Station: Go to various Cordyceps collection points to carry out mobile legal publicity activities [in Chinese]. Report 17 May. https://mp.weixin.qq.com/s?__biz=MzI4MjI4NDc5Mg==&mid=2247545375&idx=3&sn=a48b299b1837d8acb00bd20876c24076&chksm=eb9e3588dce9bc9e965b5960fb3b4f68603dbc7d8a3a3596fbbf7ac207726ca5dde4ef3faf97&scene=27

4 Trade of Raw Cordyceps

The Hui Middlemen and Tibetan Collectors Cooperatives

Jiping Sheng, Yiyuan Miao, Xinyi Xu, Lin Shen, and Ksenia Gerasimova

Introduction

The Chinese domestic trade in raw Cordyceps is based on the complex relationship between different actors, and by itself it is an interesting subject of historic and ethnographic research. The previous chapter described the painful efforts of rural inhabitants of the Qinghai-Tibetan Plateau to collect precious fungi from the grasslands. After, the raw Cordyceps pass through the hands of a middleman, who sells them to pharmaceutical companies and consumers. This chapter will give a brief overview of how a wild Cordyceps supply chain is organized in China, and then it will zoom in on the particulars of the role of Hui middlemen in this chain and the recent phenomenon of Tibetan Cordyceps collectors cooperatives, which will be illustrated with case studies.

The trade starts with drug retailers or wholesale agents from the Qinghai-Tibet area producing wild Cordyceps. Most of them are Han people or Hui people. Retail sales target tourists and local urban residents. The medicinal materials market is mostly managed by large wholesale distributors. Pharmaceutical companies have two options: to manage the supply chain themselves, including online or in store direct sales, or work with retailers.

Review of the Wild Cordyceps Supply Chain

The supply chain generally has four links: a Cordyceps collector, a primary market, a secondary market, and consumers (Figure 4.1). The primary market mainly sells the product in its original form, usually dried and unprocessed, sometimes in simple packaging. The secondary market sells refined products of high value. The products are processed to match consumer preferences, and at this stage it is possible to influence consumers' behavior by branding processed products. Depending on the trading channel, the price varies remarkably. The cost of buying Cordyceps directly from the primary collector may be several times lower than the cost at the secondary market.

The supply chain includes harvesting Cordyceps from the highland pastures, procurement and trading, processing, and retail. The different agents' involvement forms different combinations of industrial relations and affects the final price offered to consumers.

DOI: 10.4324/9781003429753-7

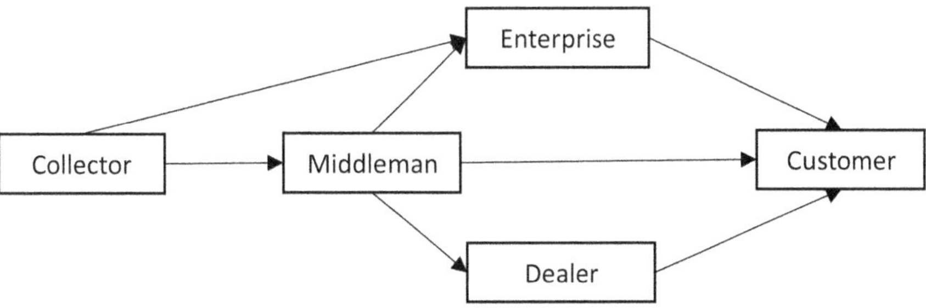

Figure 4.1 Supply chain of wild Cordyceps

Local Retailers as Mountain Contractors

Based on our observations and interviews with Cordyceps traders in Qinghai, the initial buyers of Cordyceps are local retailers, who pay for raw Cordyceps by weight, with the money going to the herdsmen who collected these Cordyceps from the Qinghai-Tibetan grasslands. The profit which Cordyceps collectors receive is small; retailers get more, but they have to bear a higher price risk. The price of Cordyceps is in constant fluctuation, and buyers are often eager to sell Cordyceps, even at lower prices. The price for Cordyceps sometimes declines due to an abundant supply or lower demand. Also, the storage period for wild Cordyceps is short. Peak collecting season is in spring/early summer, so by October the best specimens have been sold and the leftover dried wild Cordyceps reaches its minimum weight, affecting the final price.

Under this scheme, the initial retailer, sometimes referred to as a mountain contractor, signs an agreement to rent a Cordyceps pasture from a land owner and recruits farmers annually to harvest the Cordyceps. In this scenario, the profit of the retailer will be affected by two factors—land and labor. First, costs for renting land increase year on year, driven by the growing demand in Cordyceps, the falling productivity of Cordyceps-producing areas and the fixed availability of land resources. Then there are labor costs; the contractor usually employs a large number of Cordyceps collectors, who are local herdsmen and farmers, and additionally someone to supervise these hired workers and monitor the process of collecting fungi. The collectors are paid a basic salary based on the amount and quality of the collected specimen and are sometimes provided with food subsistence. All this requires big investment, and the harvest amount should be large enough to cover this outlay. The collected Cordyceps are then sold to pharmaceutical companies, who act as wholesale buyers. Thus, it is actually the mountain contractor who faces a greater market risk (fluctuating price risk) than the collectors that he hired or the Cordyceps wholesale buyers.

As explained earlier, primary Cordyceps traders either buy Cordyceps directly from herders, paying around 20–30 yuan per unit (one Cordyceps), or organize harvesting wild Cordyceps by renting pasture. Direct trade with the herdsmen cannot guarantee the quality and amount of collected fungi, so the second model of renting a Cordyceps-producing area and hiring labor, despite higher investment inputs, can secure a better quality and quantity of harvested Cordyceps. Such a model of Cordyceps supply management is mostly adopted in the Qinghai counties. As a rule, a Cordyceps mountain contractor rents Cordyceps-producing meadows on an annual basis from the herdsmen, who have the ownership rights over the grassland pastures. Usually the rent for contracted pastures, which may range from several hundred thousand to several million yuan per year, should be paid one year in advance. To rent a pasture, Cordyceps mountain contractors need to negotiate first with the local village party secretary, then sign the contract. The price ranges from 0.3 to 8 million yuan, and the grassland of about 1,000 mu might cost 8 million yuan. When the Cordyceps are to be harvested, the mountain contractor hires local or external (from other counties) workers, providing each with a basic salary and food snacks and paying about 10 yuan per Cordyceps.

The Cordyceps mountain contractor generally rents out 'collective' land from the highland village, since individual pastures are small. In other places, especially within the nature reserves, such a contracting and supply management scheme would not be possible to implement, so small scale individual Cordyceps harvesting is most common.

Hui Middlemen

Hui or Huihui literally means 'related to the Islamic faith.' The first Muslims who arrived in China were merchants and diplomats, in the 7th century. The Mongol expansion created several migration waves of Muslims from Persia to Arabia and to China, along the old trade routes known as the Silk Road. Often they came as soldiers and administrators. Over centuries they managed to maintain their ethno-religious identity and became accustomed to the general rules of the Chinese society: they 'lived on the edge of two societies and were forced to have one foot in their Islamic culture and one foot in the "host" Chinese culture' (Wang, 1996, p. 241, cited in Atwill, 2003). In fact, Islam was not just a religion but a social system that provided a unifying element within the Hui community (Gladney, 1991; Glasserman, 2021). Hui had settled in certain professions, mostly military and trade. They also enriched China by sharing access to Western scientific texts, including Arabic and translated European sources. It was Hui scholars who translated Arabic texts to Chinese (Han kitab) (Lee, 2019). Hui medicine is a unique amalgamation of Chinese, Tibetan, and Arabic medicines. There have been Hui doctors and pharmacies, and these days the interest in the history and practice of Hui medicine is growing (Zhang et al., 2023).

Historically, Hui in highland Yunnan, Qinghai and Gansu provinces were successful traders, mostly because of the poor agricultural conditions of the region and the Islamic appreciation for trade. During the Qing dynasty, which was favorable to Hui, they developed barter markets, exchanging artisan products for agricultural commodities and medicinal plants (Moevus, 1995). Close links within the closed Hui community allowed them to develop a trust-based system of lending capital and promoting trade. Frequent interactions with the Tibetans led to the phenomenon of

Diqing Hui or Zang Hui (Tibetan Hui), who borrowed from the traditional Tibetan way of life in terms of food, clothing, buildings, and language (they speak a dialect which is mix of Tibetan and Chinese) and would often marry Tibetans.

According to China's population census from 2020, the Hui is the fourth largest ethnic group in China, with around 11.4 million people, after the Han Chinese (1.29 billion), the Zhuang (19.6 million) and the Uygur (11.8 million) (Koon, 2023). Hui communities are concentrated in the Ningxia Hui Autonomous Region, Gansu and Qinghai in northwest China, and there are significant populations in Henan, Hebei, Shandong, Yunnan, and the Xinjiang Uyghur Autonomous Region (Dillon, 1994).

At one point, the Communist Party of China considered the Hui a 'backward economic element,' but with the beginning of the reform age and the introduction of the household responsibility system, the Hui's entrepreneurial skills were reassessed because they were able to increase rural earnings (McCarthy, 2009). After twenty years of trade restrictions in 1958–1978, during the reform age, Hui were allowed to trade again with Tibetans (Moevus, 1995). As many Hui speak Chinese and Tibetan, they found themselves in the favorable position of intermediary agents. Trade in Cordyceps involves several actors, as explained earlier, and it appears that Hui people have an important role, particularly in Qinghai Province.

Large conflicts with the Han Chinese over economic interests were not common, except for the Panthay Rebellion in 1856. In some places Han and Hui intermarried, and the ethnological phenomenon of 'Hui by custom' was born (Dillon, 1994).

Overall, Hui have successfully developed their own niche in the current Chinese socio-political system. Being a minority means higher mobilization and closer links within the community. In case of the Hui traders, they are very financially supportive of friends and family members, showing high trust with longer borrowing periods. From our observations, they even have their own system of hand signs to secretly communicate the acceptable price when bargaining over unprocessed Cordyceps in the market. Their sense of community, as well as good knowledge of medicine, an entrepreneurial spirit and close contact with Tibetans explains why the Hui traders occupy the role of middlemen in the Cordyceps trade.

Qinghai Cordyceps Middlemen

Xining city in the Qinghai Province is the world center of wild Cordyceps trade. Thus, we chose this location for our fieldwork. From August 6 to August 13, 2016, we conducted observations and a survey, resulting in 157 valid responses. Among those who answered, 110 (70.1%) were located in Qinfen Lane Cordyceps market (Jiuying), the largest distribution center for Cordyceps in China, 27 (17.2%) were located in Xinqian Cordyceps Science and Technology district of Xining City, and 20 (12.7%) in the Xinqian International Cordyceps World in Xining City. A second round of surveys and interviews was conducted in Qinfen Lane, May 12–13 2018.

Table 4.1 General background of Cordyceps traders

Index	*Item*	*Frequency*	*Percentage (%)*
Gender	Male	148	94.3
	Female	9	5.7
Ethnicity	Hui	133	84.7
	Han	16	10.2
	Zang	7	4.5
	Salar	1	.6
Type of registered permanent residence	Agricultural	111	70.7
	Non-agricultural	46	29.3
Marital status	Married	144	91.7
	Divorced or widowed	4	2.5
	Single	9	5.7
Age	18–29	37	23.6
	30–39	45	28.7
	40–49	46	29.3
	50–59	21	13.4
	>60	8	5.1
Educational level	Unschooled	15	9.6
	Primary school	51	32.5
	Junior high school	32	20.4
	Senior high school or Technical school	38	24.2
	College or University graduate	18	11.5
	Postgraduate or above	3	1.9

General demographic data concerning Cordyceps traders in Qinghai is shown in Table 4.1. Most of them are male (94.3%), self-identify as Hui (84.7%) and come from an agricultural household (70.7%).

Cordyceps Retail Store Operation

The primary Cordyceps retailers would usually run trading operations in their stores, selling Cordyceps and other medicinal supplies. More than half of the surveyed middlemen were running their stores as formal individual industrial and commercial households (66.2%), and a smaller group had stores registered as companies (33.8%). Most of the store owners were sole proprietors (66.9%) and the rest were joint ventures (33.1%); 65.4% of our respondents said that their business partners were their friends, and 42.3% run business with their relatives. The argument is that the Cordyceps trade is close to an ethnicity-based oligopoly. To raise funds, the store owners relied on their personal or family savings and took bank loans to start the business. When asked about access to their own Cordyceps-producing areas, only four middlemen said that they had their own pastures, and they were all Tibetan residents.

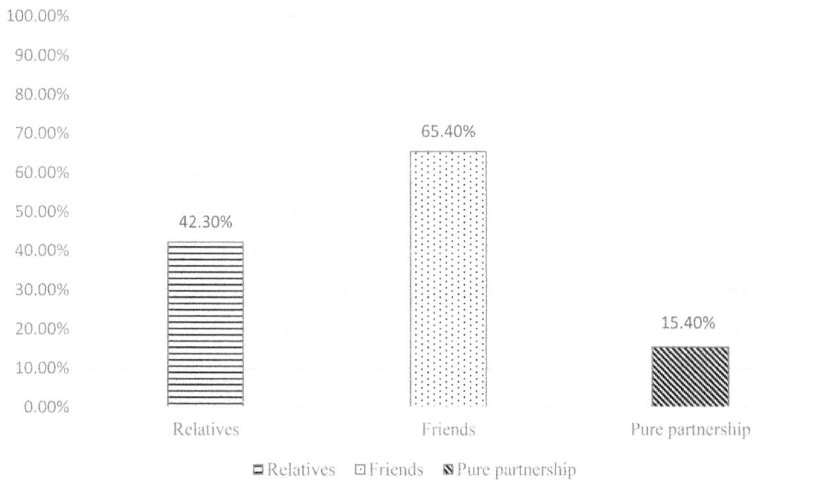

Figure 4.2 Hui business partnerships

Most of the store owners have been in business for a significant period of time. One retailer has been running his Cordyceps store for 39 years. The median period for running a store among those surveyed was 11.1 years. Overall, store owners had 1–4 stores in their retailing chains.

Buying Cordyceps

The amount of Cordyceps bought by the Qinghai traders varied from 40 kg to 3000 kg. Most traders (70.7%) bought Cordyceps from herders who harvested them from their local pastures. The remaining group of traders bought Cordyceps either from local middlemen (those living close to the Cordyceps-producing pastures) or organized collection from pastures they had access to. To purchase a large bulk of

Table 4.2 Details of Cordyceps store operation

Index	*Item*	*Frequency*	*Percentage (%)*
Type of store	Formal individual industrial and commercial households	104	66.2
	Registered companies	53	33.8
Type of property	Sole proprietorship	105	66.9
	Joint venture	52	33.1
Access to their own Cordyceps-producing pastures	Yes	4	2.5
	No	153	97.5

Table 4.3 Acquisition volume and sales credit period

Item	*Maximum*	*Minimum*	*Mean*	*SD*	*Number*
Annual volume (kg)	3,000	40	383.94	405.590	157
Sales credit period (d)	360	2	43.67	74.562	52

unprocessed Cordyceps and then sell it requires high capital input, so traders often negotiate credit with trading parties, for example with a primary retailer or a business partner.

Selling Cordyceps

The sales target for the Cordyceps traders are downstream retailers or processing enterprises. Only a very small amount of Cordyceps is sold to individual customers by Qinghai retailers. In a 2016 survey, 80.9% of the middlemen in the sales channel said that they use face-to-face sales. In 2018, the situation had already changed, as more internet sales were reported. In 2023, with the development of online shopping and the rise of live broadcasting, *Cordyceps sinensis* webcast shopping became popular, but as the authenticity of the product quality is difficult to distinguish, the regulatory norms of network sales need to be further strengthened.

In 2023, the amount of *Cordyceps sinensis* sold to all parts of the country (including Hong Kong, Macao, and Taiwan) and countries in Europe, America, Japan, and Southeast Asia was conservatively estimated to be around 180–200 tons. Domestic sales accounted for about 80% and exports accounted for about 20%. The turnover reached 30.6–34 billion yuan. The output value of Cordyceps mycelium and other derivatives can reach 57–60 billion yuan. According to Yao Xiaobao, secretary general of Qinghai Cordyceps Association, there are eight more mature Cordyceps trading markets in Qinghai Province and five more mature Cordyceps trading markets in Tibet Autonomous Region (Interview August, 2023).

As explained above, there is also a complex borrowing and lending system, especially between trusting partners, to cover high expenses before securing the final profit. As a result, half of the interviewed retailers reported that they did not own the Cordyceps they were selling in their store. The lending period varied greatly, from 2 days to 360 days, based on the trust and type of business partnership.

Detailed Case Studies of Cordyceps Middlemen

Case 1. Xinglong Cordyceps Store: Store Owner A

Initial Background

Shop owner A, a 46-year-old male of Han nationality from Gansu, has been in the Cordyceps trade for 20 years. At the time of the interview, he owned a Cordyceps specialty store in Xinqian International Cordyceps World, specializing in wholesale trade. Shop owner A managed the direct supply of Cordyceps in Yushu and Golog areas from 1997 to 2012 and then began running his store in 2013. By his estimates, the annual sales volume was nearly 1,000 kg, of which 800 kg went for wholesale and 200 kg for retail.

Cordyceps Supply

Some of his Cordyceps supplies are bought from herders who have registered access to local pastures in three locations (after May 25 in Nagqu, after May 15 in Yushu, after April 25 in Golog). Owner A also acted as a mountain contractor in two areas in Yushu and Golog, which allowed him to harvest 300 kg annually. 80% of A's stocks are purchased during the harvesting season. Owner A genuinely preferred not to go to the Qinfen Lane market to buy Cordyceps to sell in his store, citing that 'the management of Qinfen Lane [market] is more chaotic, and the quality of Cordyceps is not guaranteed.'

At the time of the interview, A had rented three 'hills' in Yushu and Golog, the largest of which was 500 mu, and he had acquired about 50–60 people to harvest the Cordyceps. Hiring such labor costs 3.5 million yuan, and they could harvest 300,000–400,000 Cordyceps during one season. Generally, collectors were paid 8 yuan for each Cordyceps, while the retail sale price per piece was 30–40 yuan on average. If the season is not productive, contractors are prone to heavy losses. Oftentimes they have to take loans, but A complained about the difficulty of borrowing in Qinghai; in Gansu Ningxia it was said that loans were more easily approved.

Cordyceps Sales

Owner A's wholesale strategy was oriented mostly towards pharmacies and drug companies in big cities, such as Zhejiang, Shanghai, Guangdong, and Beijing. Most were long-term corporate customers. The average wholesale annual trade volume was about 2,000–3000 pieces of Cordyceps.

Consumer-oriented retail was also a focus, as the store had individual loyal customers. The store was offering processed products developed from Cordyceps, such as 'Chongcao wine,' but in general, raw Cordyceps were reported to get higher sales, since the consumer cannot assess the quality of Cordyceps used in processed Cordyceps products. Owner A explained the logic:

> Take the tablets, for example, which usually contain starch. At the same time many businesses will choose poor Cordyceps for processing, and all this is opaque. Rumors say that even the leaders of the producing companies would not eat processed Cordyceps, so thousands of consumers prefer to buy raw Cordyceps just to eat.

Owner A mentioned that he never took the initiative to find new customers; they always found him. His relationships with business partners and clients were built on mutual trust. During the high harvesting season, which is 1–2 months only, old customers were allowed to sneak out some Cordyceps at wholesale price. Despite previous years' abundant supply of Cordyceps, A did not complain about the shortage.

Case 2. Zhongde Cordyceps Trading Co., Ltd.: Store Owner B

Initial Background

Owner B was 40 years old, male, and educated to junior high school level. His hometown is Huangzhong County, the largest county for producing Cordyceps in Qinghai Province, and before 2006, he worked there. After 2006, he came to trade at Qining Lane in Xining City. His Cordyceps market store had been specializing in wholesale for over 10 years. At the time of interview, his main business specialty was wholesale trade in Cordyceps, selling an approximate annual volume of about 500 kg. But he also reported that in the last few years his trade had declined, reaching only about 300 kg.

Cordyceps Supply

Store owner B said that his main channel to buy Cordyceps was through renting Cordyceps- producing pastures (the Cordyceps mountain contractor model). Additionally, if the collecting season was not providing the planned volume of Cordyceps, he would consider buying in bulk at the Cordyceps market of Qining Lane in Xining, but would first observe the market price and hunt for the best quality Cordyceps. By 2016, B had contracted two pastures, one costing 2.4 million yuan and one costing 1.0 million yuan, in his home county. The arrangement in renting 'Cordyceps mountain' involved paying half of the deposit up front and the rest at the end of the Cordyceps harvesting season. B complained that this arrangement brought him both profits and heavy losses. The price he paid to his Cordyceps collectors was 7 yuan per freshly harvested piece, regardless of size. He reported that when a Cordyceps mountain contractor hires 'someone to collect fungi,' they have to pay a deposit for each hired person as a guarantee that the grassland will not be destroyed; this money is then returned after harvesting.

Cordyceps Sales

B had a network of customers all over the country. His consumers included retail companies selling packed Cordyceps, and pharmacies, which were his long-term customers. According to B, collecting Cordyceps was easy to organize, but he feared that not all the stock would be sold. Talking about attracting new customers, B explained how new customers could be channeled from large cities, such as Jiangsu and Zhejiang in June each year.

As for Cordyceps prices, as a mountain contractor, he would buy fresh Cordyceps from hired collectors at 7 yuan per piece, as mentioned, then would sell at 30–40 per piece, but from the generated profit he also had to pay for the pasture rent. Buying Cordyceps at the Qinfen Lane market and then selling fungi to outside customers would still bring him profit, sometimes 1,000 yuan.

Case 3. Shang De Tang: Store Owner C

Initial Background

Shop owner C was 36 years old, with a college degree. He identified as Hui, originally from Xinjiang. He had been in Xining for 14 years, selling hoes to agricultural workers, and had started to trade in Cordyceps in 2013. At the time of interview, he was running a store in the Cordyceps market in Qinfen Lane. The store specialty was wholesale Cordyceps. The average annual stock in Cordyceps was 500–600 kg, providing for 95% of the trade operations of wholesale buyers.

Cordyceps Supply

C mainly purchased Cordyceps at a rural location in Qinghai County, 600 kilometers away from Xining City, from local herders, who would offer freshly collected Cordyceps to buyers at the local Cordyceps market and always bargain for a better price. C had insufficient funds to pay for labor and mountain rent. If there was a sudden demand for extra stock, C would buy around 10 kg of stock from the Cordyceps market in Qinfen Lane. If it was during a harvesting season, C would go to his rural location in Qinghai County. The most common price he reported was 2,100 pieces of Cordyceps for about 60,000 yuan.

Cordyceps Sales

C was responsible for running a Cordyceps shop in Xining, while his younger brother was also a trader, working at the market in Guangzhou. Because of this, C's customers were mainly from Southern and Eastern China and Beijing. Most buyers were long-term loyal customers, and one of them would buy nearly 200 kg of Cordyceps every year. Because of the established trust, C would often be referred to new customers, mostly downstream resellers. In fact, he had three channels for gaining new customers: existing loyal customers, his younger brother, and the Qinfen Lane market during high season. At the time of interview, C preferred traditional offline sales and had not tried online sales.

In terms of wholesale cost, 2,100 pieces of Cordyceps would cost C only 1,000–2,000 yuan. To compare, if a newcomer tried to enter the market, they would have to pay 6,000 yuan/kg. The downstream resellers would increase the price by 20,000–30,000 by the time Cordyceps reached the final consumers. C was aware that the difference in wholesale and retail price for Cordyceps sometimes rose from 40 yuan to 1,500 yuan. But he preferred to stay in the wholesale business.

In addition, due to the large sales operation and the volumes of wholesale Cordyceps sold, the customers could ask to pay on credit. But the credited sum could not exceed 200,000 yuan. This system was allowed in order to establish long-term trust, and it meant that many old customers were able to keep trading:

> And when you owe it [payment], it is convenient to talk business. The cash transaction is 70,000 per jin of Cordyceps. When you owe it, you can talk about 80,000 per jin. This is an increase in revenue for the store, but it can be paid in one month to two months.

Case 4. Zhao's Cordyceps: Shopkeeper D

Initial Background

Owner D was 43 years old, of Han nationality, and from the Tibetan village of Riyueshan, Qinghai Province, and was educated to junior high level. Before 2007, D himself collected and sold Cordyceps to middlemen, but at that time Cordyceps was only 8 yuan per piece. Then he came to Xining to trade Cordyceps and ran a store located in Xinqian Cordyceps Science and Technology City. D emphasized several times that Cordyceps was a gift from heaven to the people of the Qinghai-Tibet Plateau. His main trading product was unprocessed Cordyceps.

Cordyceps Supply

Sometimes D would go to the Cordyceps pastures to purchase stock, but most of the time family connections meant many herdsmen approached him with Cordyceps for sale. Sales were based on long-term business cooperation, trust and the credibility of business parties. In addition, D had a relative who was working as a Cordyceps mountain contractor, from whom he bought Cordyceps. This was considered a valuable connection, as the number of local Cordyceps mountain contractors was small:

> Fewer and fewer people are contracting a cordyceps mountain, as the cost of hiring people is rising and the final cordyceps output is declining. At the pastures that were previously producing 100,000 kg, it has become hardly possible to harvest 50,000 kg. In addition, as the Cordyceps mountain contractor generally pays a deposit, their investment is very high.

Cordyceps Sales

The shop specialized in selling unprocessed Cordyceps. D did not want to sell processed products, because in his opinion such products were not good quality: 'Processing a product—for example, a lozenge—requires adding a gel, otherwise it would not be of the right texture. But this is unfair to the customers.'

His main customers were pharmacies, for a relatively low wholesale price. For example, a price of 1,500 pieces/kg of Cordyceps in the Xining Lane store was 110,000 yuan, but the same amount in Tongrentang, a first-tier city, could be priced at 270,000–280,000 yuan.

Case 5. Tibetan Herdsmen Cooperatives

In the supply chain, Cordyceps collectors had bargaining power but could not always get the price they wished when they brought their Cordyceps to the market, so they often chose to work for a mountain contractor in order to receive steady payment. If a collector went on an independent hunt and sold their stock at the market, they had the bonus of being able to target tourists, who are not familiar with the price of raw Cordyceps or the standards of quality, so they could charge them a higher price (Interview with Trader, Namtso lake, May 2018).

To build bargaining power and scale up the harvested amount of Cordyceps, herdsmen can form a cooperative, which has been rather common within traditional communities based on trust and mutual assistance (Zhu, 2013).

Due to the monopoly of local middlemen at the Cordyceps market, younger Tibetan collectors reach out on social media to find customers and sell fresh Cordyceps. A young Tibetan girl, Dolma Zhuoma from Gongse Village, Daocheng County, Sichuan Province, became popular under the nickname 'Matsutake Xi Shi' on the Kuaishou social app, which she used as a platform for her videos about rural life and foraging wild mushrooms, mostly matsutake and Cordyceps. In 2017, a customer came to buy Cordyceps from Dolma's relatives' homes, and the village began to believe in her. During the 2018 Cordyceps season, Dolma used the Kuaishou platform to help the entire village sell Cordyceps and matsutake, with more than 30,000 yuan monthly revenue. In 2019, the cooperative which was set up by Dolma and local villagers earned more than 5 million yuan during the five-month picking season (Toutiao Yangtze Evening News, 2020). In harvesting and trading Cordyceps, Dolma has been involved at each stage: from digging out Cordyceps from the soil, to cleaning and drying them and then delivering to customers.

It is not easy to predict how the Cordyceps supply chain will develop in the future. But considering the diminishing supply of Cordyceps and the rising price of rent and labor, there will be fewer mountain contractors in the future, so collection will be the job of individual Tibetan herders, and they might organize themselves into cooperatives in order to get more from the trade. Hui traders might maintain the wholesale trade in Cordyceps, since they have long-established business partnerships.

References

Atwill, D.G. (2003). Blinkered visions: Islamic identity, Hui ethnicity, and the Panthay Rebellion in Southwest China, 1856–1873. *The Journal of Asian Studies*, *62*(4), 1079–1108.

Dillon, M. (1994). Muslim communities in contemporary China: The resurgence of Islam after the cultural revolution. *Journal of Islamic Studies*, *5*(1), 70–101.

Gladney, D.C. (1991). *Muslim Chinese: Ethnic nationalism in the People's Republic*. Harvard University Asia Center: Cambridge, MA.

Glasserman, A.N. (2021). *Hui nation: Islam and Muslim politics in modern China* [PhD dissertation, Columbia University].

Koon, W.K. (2023, April 21). Who are the Hui people, who lent their name to the old Chinese word for Islam and Muslims? *South China Morning Post*. https://www.scmp.com/magazines/post-magazine/short-reads/article/3217746/who-are-hui-people-chinese-word-muslim-singapore-and-malaysia

Lee, Y.L.E. (2019). Muslims as 'Hui' in Late Imperial and Republican China: A historical reconsideration of social differentiation and identity construction. *Historical Social Research /Historische Sozialforschung*, special issue: Islamicate secularities in past and present, *44*(3), 226–263.

McCarthy, S.K. (2009). *Communist multiculturalism: Ethnic revival in Southwest China.* University of Washington Press.

Moevus, C. (1995). The Chinese Hui Muslims trade in Tibetan areas. *The Tibet Journal*, *20*(3), 115–123.

Toutiao Yangtze Evening News (2020, July 1). After 95, 'Matsutake Xi Shi' Zhuoma became popular, helping villagers bring more than 300 million goods in one season [in Chinese]. https://m.thepaper.cn/baijiahao_807613

Zhang, J., Lu, L., Cai, Y., & Luo, B. (2023). Hui medicine: The Sinicized philosophical Islamic medical system. *Open Journal of Philosophy, 13*(2), 278–301.

Zhu, H. (2013). How do farmers and herdsmen participate in the market? In L. Wang (Ed.), *Breaking out of the poverty trap: Case studies from the Tibetan Plateau in Yunnan, Qinghai and Gansu*. World Scientific China.

5 Cordyceps-Based Products and Their Consumers

Jiping Sheng, Huiqi Song, Yiyun Wen, Xinyi Xu, and Shengye Shen

As Chinese people's living standards continue to rise, they focus on maintaining their health and well-being, and as a result, demand for health-related products, such as *Cordyceps sinensis*, will gradually grow. The consumption patterns of the middle class—the main consumer group—have been changing, and it is particularly important to research these trends and adapt Cordyceps-based products to meet such needs while managing natural resources sustainably. Also, the processing of *Cordyceps sinensis* should preserve its medicinal properties and medicinal efficacy, and the standardization of the products can help to improve product quality and maintain customers' safety.

Cordyceps-Based Products

Cordyceps-based products include the raw product, processed products, and fermented products. There are Cordyceps-based products that include other ingredients which can be a medicine or functional food.

Cordyceps sinensis Raw Products

Cordyceps sinensis raw products are usually Cordyceps in its natural form, unprocessed. This raw product is known as wild caterpillar fungus. As explained in previous chapters, in its raw state, *Cordyceps sinensis* has a brownish-yellow to dark brown color and a characteristic grassy or straw-like odor. It is often sold in dried form, with the moisture content being extremely low to ensure longevity and ease of storage. The processing of *Cordyceps sinensis*, to ensure the preservation and enhancement of its medicinal properties, involves several steps, which include cleaning, grading, drying, packaging, labeling and storage.

Cleaning

After being removed out of the ground, the mud on the surface of Cordyceps is retained to protect the 'placenta' from being damaged. The 'placenta' is a white film covering the surface of the Cordyceps which plays a vital role in better preserving the moisture and nutrients. The work of removing the placenta requires

DOI: 10.4324/9781003429753-8

great care and meticulousness, and it is usually done by Hui women; Hui men are involved in the purchase of Cordyceps, generally conducted at the pastures. There are two ways to clean Cordyceps.

(1) Water washing method: Sellers wash fresh Cordyceps for two reasons: first, to increase the weight of the Cordyceps; and second, to make the Cordyceps look better in color and appearance.
(2) Dry cleaning method: The term 'dry cleaning' refers to the method of cleaning fresh Cordyceps by brushing away surface impurities with a small brush instead of using water. This method is more time-consuming and labor-intensive because fresh Cordyceps are very fragile and can easily be broken if not handled carefully.

Grading

Cordyceps sinensis is graded according to its quality: premium, first grade, second grade, and third grade; the higher the quality, the higher the price.

(1) Premium Grade: The highest grade, top-quality Cordyceps. These Cordyceps are characterized by their complete and intact shape, with a lustrous black sheen and firm texture, with no damage or impurities.
(2) First Grade: These Cordyceps appear more or less intact, but there are some differences in quality when compared to the top grade. The color is slightly lighter, and the texture is softer, yet they still retain their medicinal efficacy.
(3) Second Grade: Average-quality Cordyceps that are a step down in quality from the high-grade ones. These Cordyceps might show some damage or impurities, as well as lighter color and softer texture.
(4) Third Grade: The lowest grade, low-quality Cordyceps. These Cordyceps may have significant damage or impurities, with a lighter color and softer texture. The price of low-quality Cordyceps is relatively lower.

Drying

Different drying conditions have different effects on the protein content of Cordyceps because of their various bioactive components. Protein is one of the most important nutritional components of Cordyceps. Different drying conditions may have varying effects on the protein content, thereby affecting the nutritional value of Cordyceps.

(1) Natural drying: The natural shade-drying method for Cordyceps is an environmentally friendly and traditional technique for preserving this valuable medicinal fungus. This method involves drying the Cordyceps in a well-ventilated area, away from direct sunlight. For this method, the Cordyceps is laid out on a drying surface, such as a bamboo mat or a wire screen, to ensure even drying. The drying area should be cool and well-ventilated to prevent the growth of

mold and to maintain the integrity of the product. The duration of drying can vary depending on climate conditions and the thickness of the Cordyceps; it may take several days or weeks for the fungus to completely dry out.

Once the Cordyceps is dry, it should be stored in a cool, dry, dark place to prevent moisture absorption and to extend its shelf life. The natural shade-drying method is a labor-intensive process that requires careful monitoring and adjustments to ensure optimal results. However, it is highly regarded for its ability to preserve the unique characteristics of Cordyceps, making it the best method for producing high-quality Cordyceps products.

(2) Artificial heat drying: This involves placing Cordyceps in a drying oven, where the moisture of the Cordyceps is evaporated by flowing air. Generally, the drying speed is fast and results in Cordyceps with a low moisture content. At the same time, the protein molecules are relatively well protected. Some studies have shown that heat drying Cordyceps leaves it with a higher protein content and better nutritional value.

(3) Freeze-drying technique: The Cordyceps is frozen to a temperature below the eutectic point temperature, and the water is removed by a sublimation process under a certain vacuum state. Although this drying method is expensive, the quality of the resulting product is stable and safe, which has great economic benefits and market development prospects.

Our research team conducted multiple surveys at the Cordyceps markets in Xining, Gansu Province from 2015 to 2023. We found that the most common method of drying Cordyceps for individual Hui Muslims is still shade drying. For some large companies, the drying of Cordyceps mainly involves oven drying or freeze drying. Although these artificial drying methods are more expensive, they preserve the quality of Cordyceps.

Packaging

The packaging of Cordyceps not only ensures the product's quality but also reflects its cultural significance and the care taken in its preparation for the consumer's well-being. Under normal circumstances, Cordyceps is carefully placed in packaging to ensure that it is protected from physical damage. The packaging is then sealed to prevent any contaminants and to maintain the humidity and temperature conditions that are optimal for preserving the product. There are several types of packaging for this valuable medicinal fungus: vacuum packaging; sealed packaging; glass bottle packaging; aluminum foil packaging; gift box packaging.

Labeling

Detailed labels are attached to the packaging, providing information about the product, including its grade, origin, harvest date, and any other relevant details that consumers should know.

Storage

Cordyceps sinensis is a precious traditional Chinese medicine that if stored and preserved properly can have its shelf life extended while maintaining its quality. *Cordyceps sinensis* should be kept dry, in an air-tight container, to avoid humidity and moisture. Moist conditions can cause mildew and therefore deterioration. It should be placed in a dry and well-ventilated area, away from direct sunlight and high temperatures.

Processed Products

Since ancient times, the consumption of Cordyceps has primarily been through two methods: chewing the raw fungi or preparing it as a soup. With the shift in mainstream consumer attitudes, traditional consumption methods have become too time-consuming and labor-intensive, which together with the rapid decline in Cordyceps at the same time as an increase in demand has led to the emergence of processed products. Processing can make full use of the raw material and add to its value. Cordyceps products include tablets, oral liquid, and capsules.

Chewable tablets are made by blending the powder or extract of Cordyceps with excipients and pressing them into tablet form, making it easy and convenient to consume. The tablets retain the nutritional components and bioactive substances of the original fungus, such as cordycepin, polysaccharides, and amino acids, making it an effective health supplement.

Our research group studied a Cordyceps product known as the 'Very Grass' series, produced by Qinghai Spring Medicinal Resources Technology Co., Ltd. in Xining. The series mainly consists of classic, exquisite, and raw Cordyceps, as well as double-layer tablets that use top-quality raw Cordyceps as the main ingredient. The tablets have been layered according to the natural ratio of worm and grass; the taste of 'the worm' is sweet and fresh, while 'the grass is slightly bitter.'

Fermentation Products

In toxicological and pharmacological studies, the mycelium obtained from artificial fermentation culture has been found to have the same active components and pharmacological effects of natural Cordyceps. In the production of fermented Cordyceps products on an industrial scale, the asexual stage of Cordyceps is used as a substitute for the fruiting body. This means there is low strain on the growth environment. At present, there are two culture techniques: artificial solid fermentation and liquid submerged fermentation (Wang et al., 2015).

Mycelium Powder

The scientific name of Cordyceps is Phytophthora. Since the hosting larvae are rich in physiologically active substances, Cordyceps are a valuable Chinese medicine (Xie, 2010). In the solid fermentation of Cordyceps, the main focus is the optimization of fermentation culture conditions, the optimization of solid medium components, and the detection of bacterial active components. Different from carbon, nitrogen, and temperature, the pH of the medium has a certain influence on the biomass content, and bacterial cells are the active ingredient of Cordyceps (Dong et al., 2016).

Liquid fermentation technology is quick and efficient. In recent years, liquid fermentation culture of fungal mycelium or its active ingredient, which can be used for medicinal use, has gradually been favored by the biotechnology industry and has been important for industrial fermentation production. At present, the focus is on the optimization of culture conditions and extraction methods of active ingredients.

At the early stage of industrial production, the fly lyophilized powder is used as the nitrogen source, and the Cordyceps strain is transferred to the traditional PDA oblique surface culture and then to the adaptive slant medium for adaptive breeding, and finally the production strain is produced. The compounded fermentation substratum is distributed into glass containers. These vessels are subsequently subjected to a high-temperature sterilization process to eradicate any viable microorganisms that could potentially contaminate the culture. Subsequent to sterilization, the substratum is cooled to a temperature conducive to inoculation, at which point it is inoculated with fungal spores under aseptic conditions, thereby initiating the growth of the mycelium. The inoculated containers are then incubated within an environment that has been optimized for the proliferation of the mycelium. Following an incubation period that is determined to be sufficient for optimal growth, the containers will contain a fluffy aggregation of white, dense mycelium. This mycelium is subsequently processed into a fine powder, known as mycelial powder, which can be utilized for a variety of purposes, including but not limited to dietary supplementation, bioremediation processes, or as a substratum for further cultivation endeavors (Chang et al., 2022).

Fermented *Cordyceps sinensis* powder is obtained from high-quality *Cordyceps sinensis* strains isolated and selected using modern biotechnology (Pu et al., 2023). These strains are cultivated to produce high-quality mycelium, which is then processed into a dry powder. The main components of fermented *Cordyceps sinensis* powder include Cordyceps polysaccharides, acid, amino acids, alkaloids, and other compounds. It supplements the lungs and kidneys, resolving phlegm, and relieving coughs. It can be used to alleviate symptoms such as coughing, expectoration, asthma, hemoptysis, and waist and back pain caused by deficiency of both the lungs and kidneys.

Oral Liquid

During fieldwork at the Everest Qinghai Corporation in 2015 and 2016, we obtained access to a processing line that produced Cordyceps oral liquid. The Cordyceps was planted in a potato medium at a constant temperature, and this fermentation broth was subjected to secondary fermentation for 6 days. A functional health wine of good quality was obtained, with a cordycepin content of 27.84 mg/mL and an alcohol content of 10.6% (Chen and Zheng, 2016). Cordyceps oral solution is produced using extraction technology with an active plant ingredient. Strain preparation, liquid seed preparation, fermentation, the addition of medicinal materials, ultrafine pulverization, enzymatic hydrolysis, homogenization, nano-crushing, membrane treatment, and further aseptic filling are also required. Cordyceps oral liquid was billed as a new type of product that nourishes the liver and kidneys and improves immunity and sleep.

Artificially Cultivated *Cordyceps sinensis* Products

In recent years, research on the artificial cultivation of *Cordyceps sinensis* has been conducted extensively and in-depth both in China and internationally, and although it has now been successfully cultivated, it has not been upscaled commercially.

Artificial Cultivation Techniques

Fungal Strains

Most fungal strains used are derived from natural *Cordyceps sinensis* and are isolated and cultured using conventional methods. Research indicates that during the sexual phase of the Cordyceps fungus (Wu et al., 2022), a certain amount of humidity is required for complex physiological changes, and there is a correlation between temperature and the formation of fruiting bodies. If the temperature changes slowly or remains relatively constant, it is not conducive to the formation of fruiting bodies. A low temperature and temperature fluctuation are needed during the sexual phase. After many years of research at the Sanming Fungal Research Institute in Fujian Province, only underdeveloped young stromata have been grown in culture tubes. They believe that the generation of Cordyceps must involve certain active substances from the bat moth larvae to be completed (Zhang et al., 2022).

Artificial Breeding of Bat Moths

The Sichuan Provincial Institute of Traditional Chinese Medicine has conducted long-term research on the artificial rearing of bat moths. Even under controlled temperature conditions, it takes 230–570 days to complete one generation. Moreover, the specific operations for each insect stage, as well as the feeding and rearing environment, must be handled properly. These tricky conditions are difficult for most people to create (Liu et al., 2001).

Infection Pathway of Cordyceps sinensis

This occurs when the fungus spores come into contact with the larvae, which leads to their infection and death, and then the fruiting bodies grow. Observation shows that the highest infection rate is at 4–5 instar larvae. Mature larvae are rarely infected, and larvae below the third instar are not infected. It is difficult for most people to grasp the correct timing for infection. In terms of infection, if the conditions for artificial cultivation of insects are good, the insects become too strong, with strong antibacterial resistance, making them difficult to infect (Kunhorm et al., 2019); if conditions are poor, the larvae may die after the fungus invades, leading to failure in both cases.

Simulating the Ecological Environment

This involves simulating the temperature, humidity, light, soil, vegetation, and other conditions found at altitudes of 3,500–5,000 meters, which is difficult to replicate. Claims about yields that can be harvested in a very short time are all false. Nevertheless, why do some newspapers and magazines still report on successful artificial cultivation? In fact, what they are referring to is the cultivation of *Cordyceps sinensis* mycelium or the cultivation of species related to *Cordyceps sinensis*.

***Production of* Cordyceps militaris**

Cordyceps militaris and *Cordyceps sinensis* are from the same species. Natural *Cordyceps militaris* is formed by *Cordyceps militaris* infecting insects. They are found in temperate, low-altitude habitats worldwide. In China the species is mainly distributed in Jilin, Liaoning, Shanxi, Hebei, Shaanxi, Anhui, Guangxi, Guangdong, Yunnan, Fujian, and Sichuan (Xie, 2010). *Cordyceps militaris* grows faster than *Cordyceps sinensis* and is easy to artificially culture to produce sporophore. China was the first country in the world to start artificially cultivating large quantities of *Cordyceps militaris* sporophore using insect mites, and industrial production has been achieved. In 2009, *Cordyceps militaris* was listed as a new food resource by the Ministry of Health (since renamed as a new food ingredient) (Yan et al., 2009). The main medicinal properties are close to Cordyceps in having anti-inflammatory, anti-cancer, anti-bacterial, and immune-enhancing effects, and so it can be used as a substitute. Due to limited resources of wild *Cordyceps militaris*, artificial cultivation has become popular.

Cordyceps Flower

Cordyceps flower is derived from Cordyceps. It is a new strain of fungi that has been developed by modern bio-engineering technology and can be safely consumed by people. It contains the same nutrients as those found in natural insects and is commonly included in cereals, beans, etc. It is mainly grown in the northern part of China. It is not rich in protein, amino acids, or cordycepin. In recent years, the research on Cordyceps flowers has focused on the separation and extraction of active ingredients such as total flavonoids and component content analysis (Xie, 2010). Currently Cordyceps flowers are widely sold as food in the Chinese market.

Representative Companies and Products of *Cordyceps sinensis*

Sanjiangyuan Cordyceps Technology Limited Company

The registered address of Sanjiangyuan Cordyceps Technology Limited Company is in Yushu Tibetan Autonomous Prefecture of Qinghai Province. It was established on April 9, 2002. The company's main business is the acquisition, processing, sales and research of Cordyceps and native products. The company is a leading developer and operator in the agricultural trade and the health industry. Cordyceps is the company's main product.

Qinghai Everest Insect Herbal Medicine Group

Qinghai Everest Insect Herbal Medicine Group was established in 2005 with a registered capital of 150 million yuan. Its main business is the artificial fermentation of *Cordyceps sinensis* powder. At present, the company has three main products—bailing tablets, bailing capsules, and Xueyuan soft capsules—which have obtained national standard drug identification.

Qinghai Spring Medicinal Resources Technology Co., Ltd

Qinghai Spring Medicinal Resources Technology Co., Ltd was established in 2004. It is a leading high-tech and industrialized enterprise in Qinghai Province and is headquartered in Xining. Its main products are the Very Grass series (including Cordyceps pure powder tablets, net Cordyceps and raw grass).

Bioasia Biotech and Pharmaceutical Corporation

Bioasia Biotech and Pharmaceutical Corporation is a modern biotechnology enterprise that integrates industry, academia, research, and application of medicinal and edible entomopathogenic fungi.

All the above *Cordyceps sinensis* enterprises, corresponding products, technologies, and consumers are shown in Table 5.1.

Consumption of *Cordyceps sinensis*

Current Consumption Status of Cordyceps sinensis in China

Downstream consumption of *Cordyceps sinensis* primarily based on direct consumption of the raw material stands at 94%, while processed products consumption is only at 6%. As *Cordyceps sinensis* continues to be developed towards high-tech and high-value-added industrialization, the future market capacity is expected to expand.

Table. 5.1 *Cordyceps sinensis* enterprises, corresponding products, technologies, and consumers

Enterprises	*Products*	*Technologies*	*Consumers*
Sanjiangyuan Cordyceps Technology Limited Company	Cordyceps	The company will clean sediment impurities left on the Cordyceps. They sterilize the Cordyceps and use high temperature drying; the average three pounds of wet grass is dried into one pound hay, moderately 10% humidity.	The company sells its products to national consumers through a combination of self-operated storefronts, franchisees, dealers, and online malls.
Qinghai Everest Insect Herbal Medicine Group	Artificial fermentation of *Cordyceps sinensis* powder Bailing tablets Bailing capsules Xueyuan soft capsule	The process involves isolating the *Cordyceps sinensis* from the collected fresh wild Cordyceps (scientific name: Trycoidia sinensis), after purification of strains, screening of strains, identifying traits of microbial strains of the Chinese Academy of Sciences, and authoritative identification of genetic genes, followed by the application of high-tech low-temperature deep liquid fermentation technology.	Bailing tablets and Bailing capsules are medicines. The main sales channels are pharmacies and hospitals. Products mainly enter the hospital for sale through competitive bidding. The Xueyuan soft gum and the Cordyceps fermented mycelium powder are mostly sold through health care stores or pharmacies.
Qinghai Spring Medicinal Resources Technology	Very Grass Products, including Cordyceps pure powder tablets, net Cordyceps, and raw grass	A selection of top quality raw material is used for the double layer. Smashing and solubilizing technology is used to break down the Cordyceps until it is fully dissolved. The essence obtained from this solution is not only 7 times the potency of the raw Cordyceps, but also 15% more than at the dissolution interval achieved by directly smashing the whole Cordyceps. It is then laminated according to the natural proportion of insects and grass.	The company adopts a sales model based on cooperative sales, supplemented by self-operated sales.
Bioasia Biotech and Pharmaceutical Corporation	Cicada Flower Cordyceps Series Guangdong Cordyceps Series Chrysalis Cordyceps Series Cordyceps Special Drink		The company's products have entered international markets such as Central Asia, Russia, and Northern Europe. Cicada Flower helps to meet the health needs of more elderly people.

Industrial Chain

Structure of the Cordyceps Industry Chain

The upstream industry of the *Cordyceps sinensis* sector involves the collection of this medicinal material. The industry is resource-dependent. Currently, the overexploitation and degradation of the habit in which it grows is severe, reflecting the need for sustainable management within the traditional Chinese herbal industry. The industry has developed by utilizing modern scientific and technological methods to artificially cultivate *Cordyceps sinensis*, identifying substitutes, and upscaling the value of new medicinal components. Downstream, the industry caters to the consumer healthcare market. With the improvement of living standards, people's demand for medical care is gradually increasing. At the same time, with the implementation of China's 'Belt and Road' policy, its economic influence on neighboring countries is gradually expanding. The *Cordyceps sinensis* industry, along with traditional Chinese medicine services, is gradually entering the international market. With the gradual recognition of traditional Chinese medicinal materials by pharmacopeias in Europe, America, and other countries, there is huge potential for the *Cordyceps sinensis* industry to develop further in the international market in the future.

Upstream

Within the *Cordyceps sinensis* industry, the procurement cost of the caterpillar fungus accounts for as high as 98% of all costs, while direct labor and manufacturing expenses are almost negligible. This is primarily due to the scarcity of caterpillar fungus and the increasing demand in the market, which has led to a tense supply and demand situation. *Cordyceps sinensis* can be distributed nationwide through professional medicinal material markets, making for a relatively comprehensive and smooth circulation system. The abundant stock of medicinal material resources and the well-developed market for the circulation of traditional Chinese medicinal materials provide sufficient guarantees for the development of *Cordyceps sinensis*-processing enterprises.

Downstream

Due to environmental pollution, the accelerated pace of life, work pressure, irregular diet, and lack of exercise, the physical fitness of modern Chinese people is decreasing, and a considerable part of the population is in a sub-healthy state. As people's quality of life improves and their awareness of self-care increases, more and more consumers are investing in their health. *Cordyceps sinensis* has unique nutritional functions for general health care and disease prevention, and limited resources and high market demand have kept prices high.

Cordyceps Market Policy

Before 2001, due to the absence of relevant laws, regulations, and standards for the *Cordyceps sinensis* industry, it was circulating in the market under various identities. In 2001, *Cordyceps sinensis* was listed as a national second-class protected plant, and the Ministry of Health explicitly prohibited its use as an ingredient in health products. On December 7, 2010, the General Administration of Quality Supervision, Inspection, and Quarantine officially issued a notice stating that '*Cordyceps sinensis* is strictly prohibited from being used as an ingredient in ordinary foods,' marking the end of *Cordyceps sinensis*' status as a 'food ingredient.' On August 15, 2012, the National Food and Drug Administration issued the 'Pilot Work Plan for the Use of *Cordyceps sinensis* in Health Foods,' granting *Cordyceps sinensis* 'temporary legal status' as a health product for five years. On March 4, 2016, the National Food and Drug Administration issued a notice to stop the pilot project for the use of *Cordyceps sinensis* in health foods, signaling the end of its 'temporary status' as a health product. It was not until June 2020 that the 'Pharmacopeia of the People's Republic of China (2020 Edition)' included *Cordyceps sinensis* in the 'Herbal Materials and Prepared Pieces' directory. The fluctuations in policy have had a significant impact on the entire industry, especially on the development of high-efficiency utilization and deep processing (Caplins et al., 2018).

Cordyceps sinensis Exports

While domestic demand is met, some products are mainly for export. From January to May 2023, China's volume of exported *Cordyceps sinensis* reached 0.95 tons, to the value of 13.0069 million US dollars. While the average price has been 1.36484 million US dollars per ton, the export value of Beijing's *Cordyceps sinensis* has reached 4.8778 million US dollars, ranking first in the country; Qinghai's has reached 3.6949 million US dollars, ranking second; and Jiangsu's has reached 2.2964 million US dollars, ranking third. Due to the high price of *Cordyceps sinensis*, it is mainly exported to wealthier regions (Zhang et al., 2023).

In terms of exporting destinations, China mainly exports to Hong Kong, Macau, Japan, Singapore, Thailand, and New Zealand. 'Hong Kong, China' is the largest exporting destination for *Cordyceps sinensis*, with a volume of 1,125 kilograms, accounting for 74% of exports. The next largest market is Japan, with a volume of 234 kilograms, accounting for 15% of the export market. In the future, the demand for *Cordyceps sinensis* in China is expected to expand further, with a sales value that might have already broken through 20 billion yuan, in 2023 (Zhang et al., 2023).

Current Trends in the Development of the Cordyceps Industry

As an important traditional Chinese medicine, the production of *Cordyceps sinensis* includes the following aspects:

(1) Research and application based on modern technology: With the continuous development of modern technology, the research on *Cordyceps sinensis* has become more sophisticated in terms of the isolation and extraction of effective ingredients, structural identification, and pharmacological effects.
(2) Large-scale and standardized production: In recent years, the scale of production of *Cordyceps sinensis* has gradually increased, and there is also a trend towards standardization. These measures can ensure the stability of its quality and efficacy.
(3) Diversified development and utilization: In addition to traditional Chinese medicine applications, the development and utilization of *Cordyceps sinensis* are becoming more diverse. For example, it can be used in food additives, cosmetics, and health products, and the product's time to market can be shortened, enriching consumers' choices.
(4) Development in the field of biotechnology: In the field of modern biotechnology, researchers are exploring the genetics, genes, and metabolism of *Cordyceps sinensis*, thereby promoting genetic improvement and better breeding of its germplasm resources. At the same time, based on these scientific research achievements, the application field of *Cordyceps sinensis* can be expanded by simply identifying its medicinal effects, thereby enhancing its market value.

Studies on Cordyceps sinensis Consumer Behavior

Research on Consumption

At present, there are few studies on the consumption of *Cordyceps sinensis*, and most of them are focused on the consumer's awareness of *Cordyceps sinensis* and how much it costs. He Jun et al. (2022) generated online structured surveys (n = 1,861 consumers) and semi-structured qualitative interviews (n = 65) across six provinces that comprise the primary market in China. Interestingly, consumers in higher income provinces bought less frequently but spent more money on Cordyceps for self-consumption, compared with consumers in lower income provinces, who buy more frequently, spending less, and with the intention to use purchased products as presents for family or friends. Consumers who are willing to buy healthy food on the internet account for 52.0%, and 55.1% of consumers are unwilling to believe that the product is authentic (Chen et al., 2017).

Consumers' Knowledge of Cordyceps

In order to study consumers' awareness of *Cordyceps sinensis* products, the *Cordyceps sinensis* research group of the School of Agriculture and Rural Development of Renmin University of China conducted a questionnaire survey online. A total of 310 questionnaires were completed, of which 303 were valid, with an effective rate of 97.7%. Surveyed locations were in 12 provinces and cities:

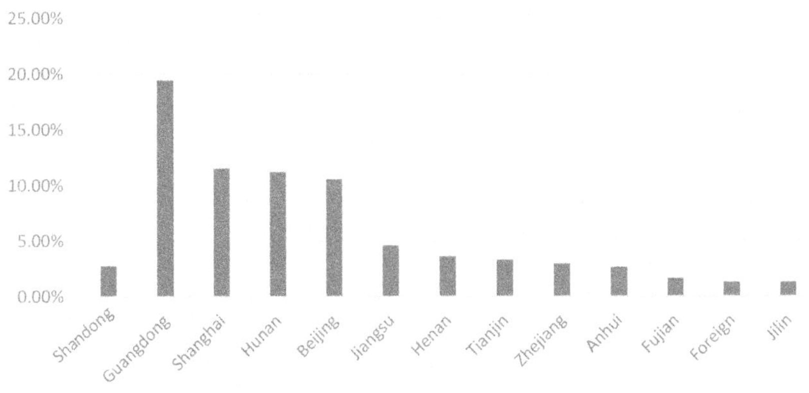

Figure 5.1 Geographical distribution sample

Beijing, Shanghai, Guangdong, Shandong, Hunan, Jiangsu, Zhejiang, Tianjin, Henan, Anhui, Jilin, and Fujian. Among them, the consumers in Beijing, Shanghai, Guangzhou, Jiangsu, Zhejiang, and Shanghai accounted for 49.2%, in addition to a few foreign consumers. See figure 5.1 for further details.

The respondents were mostly female, accounting for 62.0% of the total. The age groups were mostly young and middle-aged; young consumers accounted for 51.2% of the total sample. Customers with a junior college or undergraduate degree accounted for 63.0%, and graduate students and above accounted for 24.4%. Most people's net income is average; about 50.8% had an annual net income of more than 50,000, and 30% had a relatively high annual net income of more than 80,000.

The respondents' knowledge of *Cordyceps sinensis* is shown in Table 5.3. Around 46.9% have been exposed to information about *Cordyceps sinensis* from TV. 42.6% of people said they knew about its use, and 56.1% said they had heard of it but did not know its specific use. Meanwhile, 95.0% said they could not fully confirm their knowledge. When we asked people what they thought *Cordyceps sinensis* was used for, 40.9% of people believed that it is a health care product, and 31.0% believed that it is both a medicine and a food product. Only 20.5% of people could identify it as a traditional Chinese medicine, indicating that the attributes of *Cordyceps sinensis* have not been communicated effectively to consumers.

Consumers' Purchasing Behavior

Respondents of our survey about the consumption of the Chinese caterpillar fungus is shown in Table 5.4. Of the consumers surveyed, 19.5% have bought Cordyceps, with 12.9% buying processed products; the number buying Cordyceps for long-term use was only 5.9%.

Table 5.2 Sample of basic consumer characteristics

Indicators	*Items*	*Frequency*	*Percentage*	*Indicators*	*Items*	*Frequency*	*Percentage*
Gender	Male	115	38.0	**Health**	Fitness	218	71.9
	Female	188	62.0		Sub-health	81	26.7
Age	20–29	155	51.2		Disease	4	1.3
	30–39	41	13.5	**Elderly people in the family**	Yes	279	92.1
	40–49	42	13.9		No	24	7.9
	50–59	49	16.2	**Annual personal net income (RMB/Thousand Yuan)**			
	Below 30	109	36.0				
	Above 60	16	5.3		30–50	40	13.2
Education	Primary school and below	4	1.3		50–80	43	14.2
	Junior high school	7	2.3		80–120	57	18.8
	High school or technical secondary school	27	8.9		120–150	23	7.6
	College degree	191	63.0		150–200	11	3.6
	Graduate and above	74	24.4		Above 200	20	6.6

Table 5.3 Consumers' knowledge of *Cordyceps sinensis*

Indicators	*Items*	*Frequency*	*Percentage*	*Indicators*	*Items*	*Frequency*	*Percentage*
Where people had heard about *Cordyceps sinensis*	The newspaper	13	4.3	Knowledge of *Cordyceps sinensis*	Know/ know its purpose	129	42.6
	TV	142	46.9		Heard of it, but doesn't know what it's for	170	56.1
	Billboard	18	5.9		Never heard of it	4	1.3
	Relatives and friends	82	27.1	Can tell if something is a lie	Yes	15	5.0
	Network	48	15.8		No	288	95.0
Knowledge of the properties of *Cordyceps sinensis*	Health care product	124	40.9	Use	Care	215	71.0
	Medicine	94	31.0		Gift giving	58	19.1
	Traditional Chinese medicine	62	20.5		Cure	30	9.9
	Food	4	1.3				
	Drug	19	6.3				

Table 5.4 Sample of *Cordyceps sinensis* consumers

Indicators	*Items*	*Frequency*	*Percentage*
Have you ever purchased *Cordyceps sinensis*?	Yes	59	19.5
	No	244	80.5
Have you purchased processed products containing *Cordyceps sinensis*, such as lozenges, broken cell wall powder, capsules, and so on?	Yes	39	12.9
	No	264	87.1
Have you ever taken *Cordyceps sinensis* products over a long period?	Yes	18	5.9
	No	285	94.1

In terms of purchasing channels, consumers mainly buy *Cordyceps sinensis* in specialty stores and pharmacies. These two channels have become the most important for consumption (Figure 5.2). In terms of the main factors that consumers pay attention to when purchasing products, efficacy is ranked first followed by the brand. The specific details are provided in Figure 5.3.

Future Concerns

To provide the most recent update, we analyzed key concerns regarding future Cordyceps production as mentioned by relevant stakeholders in the Cordyceps industry whom we interviewed during our fieldwork in August 2023. These

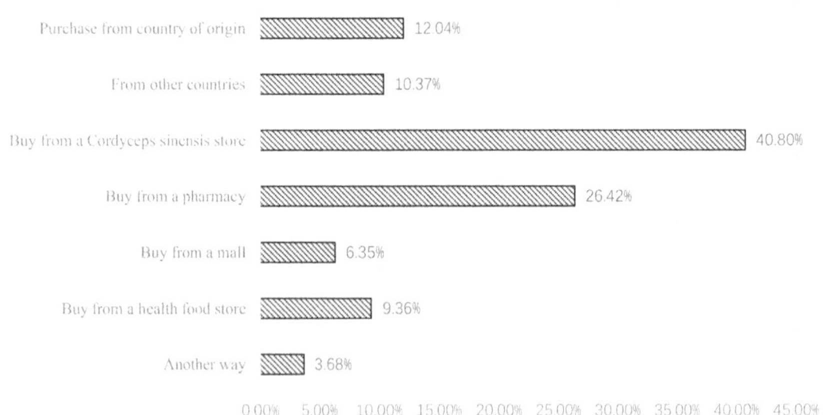

Figure 5.2 Distribution channels

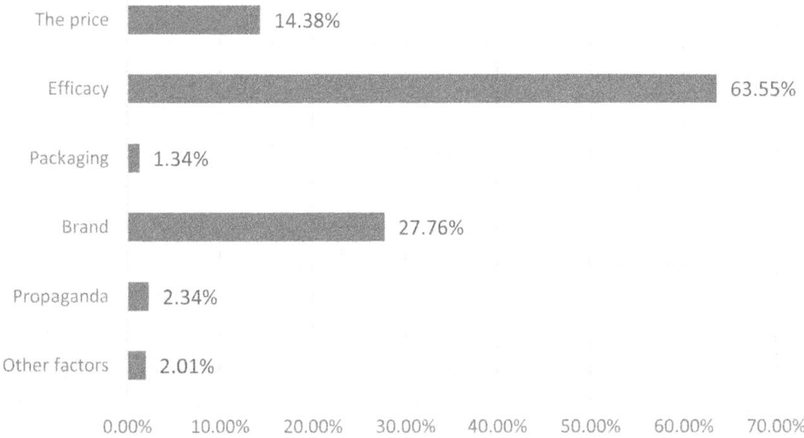

Figure 5.3 Consumer priorities

specifically include the supply of the fungi, fluctuating prices, and measures to protect Cordyceps in nature. Addressing these issues might have a stabilizing effect on the market and potentially secure the income of people employed in the industry.

References

Caplins, L., Halvorson, S.J., & Bosak, K. (2018). Beyond resistance: A political ecology of Cordyceps as alpine niche product in the Garhwal, Indian Himalaya. *Geoforum*, *96*(11), 298–308.

Chang, H.H., Chang, W.J., Jhou, B.Y., Kuo, S.Y., Hsu, J.H., Chen, Y.L., & Lin, D.P.C. (2022). Efficacy of Cordyceps Cicadae (Ascomycota) Mycelium supplementation for amelioration of dry eye symptoms: A randomized, double-blind clinical pilot study. *International Journal of Medicinal Mushrooms*, *24*(12).

Chen, X., Sheng, J., & Wang, H. (2017). Optimization of CS. SYSU-II spray drying process of *Cordyceps sinensis* by response surface methodology. *Food Industry*, *38*(8), 117–112.

Chen, Y., & Zheng, L.V. (2016). Advances in the extraction of cordycepin and its application in the beverage industry. *Beverage Industry*, *19*(1), 50–52.

Dong, C., Li, W., & Li, Z. (2016). China's Cordyceps industry development status, problems and prospects—Jinhu Declaration on Cordyceps Industry Development. *Journal of Fungi*, *35*(1), 1–15.

He, J., Smith-Hall, C., & Zhou, W. (2022). Uncovering caterpillar fungus (Ophio*Cordyceps sinensis*) consumption patterns and linking them to conservation interventions. *Conservation Science and Practice*, *4*(8), e12759.

Kunhorm, P., Chaicharoenaudomrung, N., & Noisa, P. (2019). Enrichment of cordycepin for cosmeceutical applications: Culture systems and strategies. *Applied Microbiology and Biotechnology*, 103, 1681–1691.

Liu, Z.Y., Liang, Z.Q., Whalley, A.J.S., Liu, A.Y., & Yao, Y.J. (2001). A new species of Beauveria, the anamorph of Cordyceps sobolifera. *Fungal Diversity*, 7, 61–70.

Pu, S., Yang, Z., Zhang, X., Li, M., Han, N., Yang, X., & Shao, H. (2023). Fermented Cordyceps powder alleviates silica-induced pulmonary inflammation and fibrosis in rats by regulating the Th immune response. *Chinese Medicine*, *18*(1), 131.

Wang, X., Lin, S., & Liu, Z. (2015). Research progress on artificial culture of *Cordyceps sinensis*. *Genomics and Applied Biology*, *34*(7), 1569–1574.

Wu, H., Liu, H.N., Ma, A.M., Zhou, J.Z., & Xia, X.D. (2022). Synergetic effects of Lactobacillus plantarum and Rhizopus oryzae on physicochemical, nutritional and antioxidant properties of whole-grain oats (Avena sativa L.) during solid-state fermentation. *Lwt*, 154, 112687.

Xie, L. (2010). The difference between Cordyceps flower and *Cordyceps sinensis*. *Strait Pharmacy*, *22*(7), 54–55.

Yan, J.K., Li, L., Wang, Z.M., Lueng, P.H., Wang, W.Q., & Wu, J.Y. (2009). Acidic degradation and enhanced antioxidant activities of exopolysaccharides from *Cordyceps sinensis* mycelial culture. *Food Chemistry*, *117*(4), 641–646.

Zhang, Z., Chen, W., Liang, J., Zhang, L., Han, Y., Huang, J., & Liang, Z. (2022). Revealing the non-overlapping characteristics between original centers and genetic diversity of Purpureocillium lilacinum. *Fungal Ecology*, 60, 101179.

Zhang, H., Yang, J., Luo, S., Liu, L., Yang, G., Gao, B., & Yang, M. (2023). A novel complementary pathway of cordycepin biosynthesis in Cordyceps militaris. *International Microbiology*, 1–13.

6 International Dimensions of the Chinese Trade in *Cordyceps sinensis*

Ksenia Gerasimova, Ekaterina Zueva, and Jiping Sheng

In this chapter we discuss historic developments and key trends in international markets for Cordyceps. It will start with a brief chronology and description of the internationalization of Traditional Chinese Medicine (TCM) then will briefly review supply chains of *Cordyceps sinensis* outside China. *Cordyceps sinensis* is commonly referred to as yartsa gunbu, which translates from Tibetan as 'winter worm, summer grass,' and in Himalayan it is called keera jadi, referring to the larva (keera) and the emergent fruiting body that resembles sprouting grass. It is endemic to the Qinghai-Tibet Plateau and grows at altitudes of 3200–4900 m, with a non-vegetative cover of 15–55% and an average winter temperature of -15 to -5°C. Thus, based on these conditions, only Bhutan, India, and Nepal, in addition to China, can produce wild *Cordyceps sinensis*. For centuries, Tibetans collected Cordyceps for their own use and sold it to Hui and Han traders. Once the Chinese domestic market was established, neighboring countries were also incorporated into the supply chain. All four *Cordyceps sinensis*-producing countries face similar issues: high reliance on rural residents, particularly the transhumant minorities, to collect the caterpillar fungi; overharvesting, and the need to regulate access to natural resources and protect this endangered species. The Cordyceps-based products are mostly targeted at customers in the Asia-Pacific region and the USA, while customers in Europe and the rest of the world have also shown interest in Cordyceps. To illustrate the development of new markets for Chinese Cordyceps, the Russian market is an unusual but interesting example.

The Internationalization of TCM and the Global Export of its Associated Products

Cordyceps sinensis is known as one of the ancient treasures of Traditional Chinese Medicine (TCM). TCM was largely unknown to the outside world until the period of Great Discoveries and European colonization. In his *Travels*, Marco Polo introduced his European readers to new lands, including China, and among other mysterious products referred to traditional Asiatic medicines, such as rhubarb, ginger, and snake's gallbladder (Polo, 1845, pp. 83, 163). But most knowledge was kept within China.

DOI: 10.4324/9781003429753-9

In the 20th century, following the strategy of creating New China, the Communist leaders chose, for the country's medical services, a strategy of combining TCM with Western science. While some argued that the policy of combining two medicines, also called 'walking on two legs,' was designed out of necessity, as no other medical facilities were available, 'the Westernalisation of native medicine' was considered compatible with Communist science in the Chinese Communist ideology (Croizier, 1968). Foreign experts were invited to learn more about the foundations of TCM and to share Western scientific practices. These visits became the first points of scientific communication and knowledge exchange. For example, after such a visit to China, a British professor in anatomy wrote an article for *The Lancet* about TCM, inviting further investigation (James, 1955). The key focus of the Communist modernization of TCM was the public spread of knowledge of medical practices, which had been kept private for centuries by TCM practitioners. The ancient science was acknowledged but also modernized and institutionalized, with the first TCM colleges and medical conferences organized especially after 1955 (Croizier, 1968).

Xu et al. (2013) offer a chronology of the TCM modernization process in China:

Phase I (1950s–1970s): developing TCM higher education, research and hospital networks in China;

Phase II (1980s–2000s): developing legal, economic and scientific foundations and international networks for TCM;

Phase III (2011 onwards): further consolidating the scientific basis and clinical practice of TCM. Starting from 2010, around 255 new investigational drugs were derived from TCM herbs.

The current TCM market is based on practices such as acupuncture, cupping therapy, moxibustion, aromatherapy, compounding therapy, and magneto therapy (PMR, 2020). TCM has studied the medicinal effects of 11,000 herbal plants and most commonly uses 700 species as raw materials (Xu et al., 2013). Chinese Customs reported that the country has been exporting 240,000 tons of medicines annually, of which 200,000 tons are raw herbs (20% of the country's annual harvest). As a result, TCM plant production has become a new source of income for Chinese farmers. Provinces such as Hebei, Guizhou, Yunnan, Sichuan, and Shanxi have been identified as key bases for TCM production (ibid.).

Market development is usually defined by an increase in market size, product value and volume of sales, recorded over a period of time. Economists use CAGR or compound annual growth rate to calculate the rate of the value change in the market. Usually, the price reflects the demand for a product. Based on the price dynamics recorded by Winkler (2009), Cordyceps has become the most valuable biological commodity in the world. There are several reasons for this. First, TCM, including *Cordyceps sinensis*, has gained more recognition outside China due to globalization and expanding international trade. The CAGR of TCM in the period 2003–2011 was 24%. Its value reached USD 14.135 million by 2014. The main importing countries were in Asia Pacific, which accounted for almost 45% of all exports (Mucelli, 2018). This was the 'golden age' for TCM trade.

In the period 2014–2017, TCM exports volume dropped, with the CAGR in 2014 reaching 17.89%, and the TCM market value at 2.557 billion in 2017 (ibid.). However, in 2015 China significantly improved its centrality within the global trade network; in effect, it meant that more countries had started to export TCM products. The key developments included more stringent regulations regarding manufacturing techniques related to TCM drugs and optimization of production processes and minimization of contaminants to comply with pharmacopeia standards, such as USP, BP, EU and IP (ibid.).

Another reported trend is the rising acceptance of TCM. This is based on people's increased interest in treatments, with the global TCM therapy market witnessing rising demand. Lack of specific guidelines for manufacturing and approval of TCM is deterring research and development initiatives related to novel formulations in CNS diseases. Under the present regulatory scenario, if a new drug approval (NDA) fails, the herbal extract and API also fails even if it is not due to anything related to it or due to excipients (PMR, 2020).

Development of the International Market for *Cordyceps sinensis*

Cordyceps sinensis was classified as a drug of TCM in 1964 by the Chinese state (Paterson, 2008). In 1993 the Chinese running team under the coach Ma Junren was reported to have broken three world records after taking supplements based on Cordyceps (Hooper, 2005). During the severe acute respiratory syndrome (SARS) epidemic in 2003, the price of *Cordyceps sinensis*, which was regarded as the protective treatment against SARS, skyrocketed, exceeding around 30,000 RMB/kg and up to 100,000 RMB/kg in 2006 (Dong et al., 2015). *Cordyceps sinensis*, as shown in Chapter 2, was used as a complementary therapy to treat the COVID-19 epidemic in China, and this inspired substantial research globally.

It is very difficult to estimate the annual global supply of *Cordyceps sinensis*; the numbers may be around 185 tons per year (Elkhateeb & Daba, 2020). Based on the interviews conducted with Cordyceps traders in China and Bhutan, and the published data from China (Table 6.1), we estimate this number to be around 210 tons. Since

Table 6.1 The supply and demand of wild Cordyceps

Time (Year)	*Production Volume (Ton)*	*Demand Volume (Ton)*	*Difference (Ton)*	*Growth Rate*
2014	93			
2015	125			34.40%
2016	109	108.4	-0.6	-12.80%
2017	224.5	223.5	-1	106.00%
2018	175.9	175	-0.9	-21.80%
2019	192.3	191	-1.3	9.40%
2020	159.2	224.3	65.1	-17.20%
2021	192.6	244.2	51.6	20.98%

Source: Huaon Research Institute (2021).

natural Cordyceps production is highly dependent on the conditions of its habitat, such as winter and summer temperatures, volume of snow, soil precipitation and intensity of sunlight, the annual production fluctuates. At the same time, consumer demand significantly increased in 2020 and continued to grow in 2021–2022.

The most expensive price for *Cordyceps sinensis* we recorded was in 2022–135 (¥/g) at the Chinese online platform Taobao. To contrast, in the early 1970s the price was only 20 ¥/kg (Winkler, 2009). It is indeed the most expensive commodity in the world with an almost 700% increase in price.

In 2022 the global market value of *Cordyceps sinensis* was estimated at 1.12 bn. USD in 2023, with a projected growth of 1.60 billion USD in 2027 at a CAGR of 9.4%. A relative of *Cordyceps sinensis*, *Cordyceps militaris* also presents useful medicinal qualities and is widely used around the world. But for commercial production, artificial cultivation is most common (Li et al., 2019; Ren et al., 2021). Thus, the market value of *Cordyceps militaris* was estimated at 1.62 bn. USD in 2023 and has been expected to reach 3.11 bn. USD by 2032. While the *Cordyceps sinensis* market is dominated by Chinese companies, American companies prevail alongside the two Chinese companies in the *Cordyceps militaris* market (PMR, 2020; VMR, 2023). The key market players are listed in Table 6.2.

Bhutan

Bhutan, a small kingdom hidden in the Himalayas, has also experienced 'the Cordyceps rush.' Ordinary people used to collect fungi for personal use, but in the 21st century they started to collect fungi for commercial trade and sell it to Chinese and Indian traders. Rural Bhutanese also wanted to benefit from high prices, but the local variety of *Cordyceps sinensis* is slightly smaller and lighter than the Chinese variety (Interview with a park guide, Lam Perli, March 2015). Effectively, it means that the Bhutanese collectors need to spend more time harvesting Cordyceps than their Tibetan counterparts in China.

The collection of Cordyceps was officially prohibited under the Kingdom's Forestry Act until 2004. First, harvesting in one area—Lunana district—was legalized then more areas were added to the list after 2004. According to the regulations, the right to collect precious fungi was granted only to highlanders

Table 6.2 Key players in the *Cordyceps sinensis* and *Cordyceps militaris* markets

Cordyceps sinensis *Market*	Cordyceps militaris *Market*
Lei Yun Shang Pharmaceutical Group	Aloha Medicals
Tongqingyutang	Nammex
Shenxiang	Mushroom Science
Sanjiangyuan	Naturalin Bio
Zhufengshengao	Nutra Green Technology
Kang Mei	Qingdao Dacon Trading
Huqinggyutang	Xian Saiyang Bio-Tech
Jinkezangyao	Health Essence
	StarWest Botanicals

living in Gasa, Bumthang, Thimphu, Paro, and Trongsa districts with the intention to earn additional income (Dorji, 2023).

Wangdue-Phodrang is reported as Bhutan's most productive administrative unit for Cordyceps collection (Thapa, 2017). Pastures on the slopes of the highest mountain of the country—Mt Gangkhar Puensum, the tallest unclimbed mountain in the world—have also been popular with Cordyceps harvesters. To be able to collect Cordyceps, local herders need to have been registered in the gewog (local administration) and use their highland pastures (tsamdros) for raising yaks, and absentee yak herders, who own access to pastures and raise yaks but moved to live in lowland valleys, are also considered eligible to harvest Cordyceps (Lhamo, 2022).

Despite the system of strictly regulated short-term permits, illegal gatherings are common. In 2023 around 200 illegal collectors were caught in five different locations in Lunana district. Forest guards, soldiers and representatives of the local committees joined forces to stop illegal harvesting (ibid.).

Bhutan has an established system of natural parks and reserves. During the Cordyceps season, most of the employees are called to protect the parks from intruders (Interview with a park guard, March 2015). Another measure to stop illegal harvesting was to regulate the Cordyceps auctions. The government imposes a 4.9% levy on sales at auction to support environmental protection programs (Cannon et al., 2009).

Alongside came a grassroots initiative—a group of 25 yak herders from 5 different Cordyceps-producing valleys, namely Soi and Lingshi (Thimphu), Laya and Lunana (Gasa) and Dur (Bumthang), set up an informal organization to control local trade and prices, making sure that harvested Cordyceps bypass illegal international traders and address local needs first—local exporters and hospitals (Tshetem, 2019).

Bhutan's approach to managing the yartsa gunbu harvest sustainably included: relaxing the laws on gathering yartsa gunbu in order to provide local people with an incentive to police their areas and protect natural resources; restricting the number of harvesters to a few members per household; establishing a law that yartsa gunbu can only be sold at authorized auctions by authorized collectors and that buyers must be Bhutanese nationals (Cannon et al., 2009). But the Bhutanese herders preferred markets to auctions, since the bidding prices sometimes do not hit the target the herders are satisfied with, and sometimes they pulled out from the auction; illegal harvesting is too common (Palden, 2017).

Signs of the overharvesting and natural degradation have manifested even in Bhutan, the most pristine country in the world. In some parts of the country, there has been degradation of shrub plantations near the tree line, especially old-growth rhododendron plantations, due to seasonal Cordyceps harvesters' reliance on fuelwood (Wangchuk & Wangdi, 2015). Herders now face a serious challenge. While collecting Cordyceps gave them an annual income (42,000 USD) which is much higher than the average income from animal raising (155 USD), future harvesting has been endangered by the overharvesting and rising temperatures. Bhutanese herders have started to consider going back to raising yaks as a substitute for collecting Cordyceps:

> There are chances that cordyceps will disappear one day. At such a time, the highlanders might suffer and will have to depend on their livestock once again. But it will be a difficult task for everyone to revive the tradition of rearing yaks and other animals then. Before this fungus vanishes, I urge everyone to rear at least a few yaks. The government and the herders should work together to protect these animals into the future.
>
> (Tauchu Ragpay, cited in Thapa, 2017)

India

The Himalayan region of India is the main source of a wide range of non-timber forest products (NTFPs). The main products are traditional medicinal herbs and essential materials for the aromatherapy oil industry in India. Cordyceps can be found in the highland Himalayan pastures in the states of Uttarakhand, Sikkim, and Arunachal Pradesh. Many of these areas are protected, such as Kanchendzonga Biosphere Reserve in Sikkim, Dehang-Debang Biosphere Reserve in Arunachal Pradesh, Nanda Devi Biosphere Reserve and Askot Wildlife Sanctuary in Uttarakhand (Negi et al., 2020). *Cordyceps sinensis* covers about 72 km (1.01% of the total geographical area, 7096 km) of Sikkim—confined to the North (56 km) and East districts (16 km).[1]

While the folk medicines Ayurveda and Amchi have utilized Cordyceps for commercial products, the fungus was rediscovered in 2001 or 2003 when Sikkam locals learnt about the rising prices in the international market from traders from Terai (the lowland neighboring region in Northern India and Southern Nepal in the outer foothills of the Himalaya) (Negi et al., 2006; Moudgil, 2019). Before, in the early 1990s, informal trade of Cordyceps in Sikkim was initiated by herders from Nepal, who had been hired for yak herding in Lachung. They had prior knowledge about the international market, as Cordyceps had been traded in Nepal since the late 1980s. Later on, the Lachungpas took over the trade. In Lachen, its collection and trade started after the area opened for tourism in 2005, while in Gnathang, its popularity increased after the reopening of the Nathula trade route in 2006 (Pradhan, 2016).

In the Uttarakhand highlands, Bhotia (or Bhotya) people are the main collectors of *Cordyceps sinensis*. They are a caste of shepherds of Indo-Tibetan origin, divided into seven tribal groups. 'Bod' was the Indian name for Tibet, and Bhotia used to live alongside the 'Bhot tract,' part of the Silk Road route between India and Tibet. The majority follow Hinduism and celebrate traditional Indian festivals such as Druga puja and Diwali (Sharma, 2020). They are also transhumant herders. They share a social and cultural destiny with Tibetans in TAR and Qinghai, China: they have lost access to some of their lands due to conservation projects in the Indian Himalaya and growing infrastructure, thus collecting Cordyceps became a substitute income which did not require initial capital investment or education. No wonder Cordyceps have been overharvested when whole villages, including former residents who now live in cities, rush to the highland pastures (Moudgil, 2019). Similar to Tibetans, the Bhotia collectors experience environmental and social hazards while harvesting precious fungi: highland and river passages are dangerous to climb and disputes over the collected harvest are common. To reduce these risks, Indian collectors usually camp together in pastures to collect Cordyceps in groups, relying on a system of mutual help (Caplins & Halvorson, 2017). Collectors in Lachen and Lachung usually went as a group of 5–10 people and would then spread across the slope to ensure maximum collection of Cordyceps and also prevent damage to the habitat and Cordyceps stroma through overcrowding, trampling, and soil compaction. The local collectors made sure that holes were filled with soil after digging out Cordyceps. The collectors in Gnathang collected Cordyceps opportunistically on a daily basis and had done so for decades (Pradhan, 2016; Pradhan et al., 2020).

Indian Cordyceps are usually collected from the second week of May to the first week of August. The collectors start foraging early in the morning and continue as long as the weather permits or until the evening. When Cordyceps are found, collectors remove the soil with a stick or a shovel, and, holding the caterpillar firmly, pull it out along with the soil without breaking the piece. After, collected Cordyceps are gently cleaned with a toothbrush and left to air dry; collectors return to the village or work in turns from their camps (Moudgil, 2019).

The wet climate presents a challenge for drying Cordyceps, as mold quickly destroys the precious harvest. On reaching the village, the collectors rush to sell their harvest to the sublocal trader on a per-piece basis. The trader thoroughly cleans and dries them further at room temperature by spreading them out on a newspaper. The dried materials are then wrapped in tissue paper or a muslin cloth and stored in an airtight container until they are sold (Moudgil, 2019; Pradhan et al., 2020). In Gnathang, the cleaned materials, as a local practice, were wrapped in muslin cloth and hung over the bhukari (room heater) for drying and storage. Some collectors dried them by spreading them directly over bhukari before storing them in an airtight container for future use or sale (Pradhan 2016; Pradhan et al., 2020).

Land rights have resulted in overharvesting 'keeda jadi.' Since there is no caste or class barriers in harvesting Cordyceps, many outsiders have come to collect the fungi and sometimes have left very little Cordyceps specimens for reproduction, leading to a 15% fall in the amount of produce (Moudgil, 2019). Highland Cordyceps-producing pastures have been managed by the villagers and by the state (forest guards). Commercial harvesting of yartsa gunbu was legal only in community forests managed by van panchayats (Community Forest Councils) with the approval of the State Forestry Department (Wallrapp et al., 2019). While the villagers zealously protect their lands from outsiders, forest guards' teams have been understaffed and sometimes even abused their power to take freshly harvested Cordyceps from villagers (Moudgil, 2019). As a result, collectors collect yartsa gunbu wherever they can—in state forests or in protected areas—regardless of ownership rights and legal status (Negi et al., 2015). Dr C.S. Negi attempted to explain to the villagers the necessity of limiting access to Cordyceps-producing pastures by issuing local permits, but the villagers refused, being afraid to lose their income (Pradhan et al., 2020).

In addition to the State Department of Forestry, new institutions of natural resource management have emerged at the state and local levels in recent years: the State Council for Biodiversity and the Biodiversity Management Committees (BMCs). Some functions of the BMC overlap with the rights of the van panchayats, especially with regard to regulating community members' access to resources. BMCs have been legalized in accordance with the National Biodiversity Conservation Act (2002). In the Cordyceps-producing regions, 14 BMCs have been created, but they are not yet fully operational, not fully accepted by society, and not integrated into existing local natural resource management systems. Currently, the van panchayats are having a greater impact. At the local level, leaders of van panchayats are better known, and their decisions and actions are legitimized by community members. However, in the future, van panchayats may be replaced by BMCs in communities if BMCs receive more financial resources and government support and therefore accumulate more authority and power (Pauls & Franz, 2013; Singh 2016).

The supply chain of Cordyceps in India is organized in a similar manner to China. Medicinal and other products are sold by collectors to intermediary traders, who then resell the fresh Cordyceps to wholesale traders. Through this chain, most Cordyceps are supplied to Indian processing companies (Interview with Cordyceps trader, Dehradun, March 2015).

Nepal

In Nepal there are 26 districts where, at the altitude of 3,000–5,000 m, *Cordyceps sinensis* is harvested. Out of the 13 districts involved in commercial harvesting, Bajhang, Darchula, Dolpa, Jumla, and Manang contribute to 95% of total commercial trade (Paudel, 2023).

National environmental legislation imposed in 1973 created a system of nature reserves, including those in the Nepalese Terai, to protect threatened species (Heinen & Yonzon, 1994). In accordance with Nepal's national policy, the collection and sale of NTFP, including yartsa gumbu, is legal, but the total annual amount that can be collected from a protected area is limited by the Department of National Parks and Wildlife Conservation (DNPWC).

In the 1990s, over 70% of the Nepalese population relied on agriculture for their livelihoods, and cases of malnourishment and hunger were common. The natural environment has been an important source of food for the local population and sustainable management of environmental resources was not effective. In some rural areas, the cash received for the NTFP harvest is the only income and can contribute to more than 50% of the annual average income of the Nepalese population (Edwards, 1996). In the Cordyceps-producing areas, collecting and selling caterpillar fungi contributes to 70–87% of rural households' income (Paudel, 2023).

It has been reported that some households specialized in collecting medicinal plants to the extent that they gave up other livelihood activities, such as crop production, as the Cordyceps harvest season coincides with the agricultural season. However, there might have been several other reasons for this other than farmers wanting to concentrate on Cordyceps, such as climatic changes (Shrestha & Bawa, 2013).

Despite the Forest Act 1993 and Forest Regulations 1995, villagers, including forest rangers, were not aware of what species and in what quantities they were allowed to forage (Adhikari et al., 2016). As for harvesting *Cordyceps sinensis*, it was ineffectively banned till 2000 under these regulations. *C. sinensis* harvesting began on a large scale in 1998, and the first tractor drove to the highland area in 2012 (Timmerman & Smith-Hall, 2020).

Realizing its inefficiency, in 2001 the Nepalese authority lifted the ban on Cordyceps collection, with the provision of revenue of NRs 20,000 per kg, which was later reduced to NRs 10,000 per kg, in 2006 (Devkota, 2010). With rising prices for Cordyceps, many outsiders left the Cordyceps-producing pastures. This created conflicts among the residents of the lowlands and the Shauka highlander community. Shaukas are an ethnic tribe in the Central Himalayas, seasonal herders and barter traders who speak a Tibetan dialect and are related to Kumaun and follow Buddhism and Hinduism (Bhakuni & Chaunan, 2023). They lost some of their income when the trans-Tibetan trade was disrupted (Negi, 2017). Shaukas and Rang, who had reduced their number of Cordyceps collectors for several reasons, including climatic and environmental conditions, united to protect their pastures from outsiders, issuing informal permits and limiting access to one small Cordyceps-producing area, arguing that the fungus was not a daily essential, such as fuelwood and fodder. In return, the lowland villagers, organized by the Village Development Committees, closed access to winter pastures. The conflicts in the Api Nampa Conservation Area, which has annually produced hundreds of kilograms of yartsa gunbu, have led to several killings (Pant et al., 2017).

Local institutions have attempted to regulate and promote sustainable practices. For example, in Samagaun, Nubri Valley, the Village Development Board has developed and implemented the Yartsa Gunbu management regime (Negi et al., 2015). The community leaders set the harvest start date, and every able-bodied resident was expected to register at the community meeting house. Anyone caught visiting Cordyceps pastures before the start date was subject to a heavy fine. Only conscientious villagers, whose status is determined by participation in the household taxation system, have the right to collect yartsa gunbu. Each registered household must pay a Yartsa Gunbu tax of 100 rupees for the first member of the household and then more for each additional member; the money collected in this way is spent on social events such as inviting a lama to an empowerment ritual (dbang) (Childs & Choedup, 2014; Negi, 2017).

A partial solution to reduce overharvesting has come from religious practices. In Nubri and Tsum, religious decrees (chos khrims) prohibit collection in certain sacred areas, ensuring that parts of the landscape remain undisturbed. Lamas issued 'sealing decrees' (shag rgya) declaring that tracts of land on the slopes of Gang Pungyen (Mount Manaslu, 8,156 m, the abode of a local deity) were sacred and prohibiting locals from cutting trees, gathering forest products (including yartsa gunbu), or hunting wildlife (Negi et al., 2015).

The Russian Cordyceps Market

Russia has been following Western science to treat medical conditions, although it has still been common for Russians to use medicinal herbs they harvested themselves, and this traditional background explains why Russian consumers could be more open to buying Cordyceps. Indeed, it is a growing market for *Cordyceps sinensis* and *Cordyceps militaris* that requires further regulation.

Wild Cordyceps can actually be found in the Asiatic part of Russia, especially Siberia and the far east. Due to how difficult it is to access (harsh climate and impassable swamps) along with fluctuations in growth, wild Cordyceps (Cordyceps Fries, *Cordyceps militaris*) have not been comprehensively studied in Russia, but they are included in the list of protected species (the Red Book of species). In Eastern Siberia, the largest population of *Cordyceps militaris* is found in the Khamar-Daban mountains, near Vydrino, Mamai, Tankhoy stations, Buryatia. Cordyceps are also found in the nature reserve of Baikal Lake. Isolated areas of Cordyceps habitat include the Krasnoyarsk Territory and Irkutsk and Chita regions (Ministry of environmental resources of Buryatia, 2023). *Cordyceps militaris* has also been found sporadically in the Novosibirsk and the Krasnoyarsk regions (Lednev et al., 2007). Cordyceps are found in taiga, mountain taiga, as well as tundra and forest tundra, but these places always have high levels of moisture in the soil. Floodplain forests and, less often, boreal thickets of deutzia shrub provide suitable conditions for cordyceps. In Siberia, Cordyceps usually attack larvae and pupae of various Lepidoptera, and less often imago Diptera (Ministry of environmental resources of Buryatia, 2023).

While it is common for Russian rural residents to harvest wild mushrooms, Cordyceps are a rare specialty. We found a trader of Cordyceps in the Sunday market in Tyumen, Southern Siberia. This 65-year-old woman specialized in collecting and processing non-timber forest products from Siberian forests and claimed to sell harvested Siberian Cordyceps (Interview September 8, 2023, Tyumen, Russia). During our observations at the market, we asked visitors about their knowledge of Cordyceps. Only two people out of 50 interviewed had heard of them.

The Russian-Ukrainian War and the economic sanctions imposed on Russia have greatly affected the Russian pharmaceutical market and prognosis of oncological patients. Many doctors have been afraid to speak in public, but according to the director of Federal Science Clinical Centre of Hematology and Oncology, Professor P. Trahtman, the sanctions have led to delays in equipment and drugs to treat cancer; tamoxifen has become unavailable, and he estimated arise in deaths among his patients (Filina-Kogan, 2022).

Such a background explains why Russian oncological patients and those with serious chronic conditions have turned to traditional medicine, while market development has been bolstered by the arrival of lab-grown Cordyceps-based products. We found an informal online forum for oncological patients and their families, who were sharing their experiences in treating different types of cancer with Cordyceps. Individual traders, who keep their identity private, supplied their raw *Cordyceps sinensis* from TAR, Tibet. Many patients have used *Cordyceps sinensis* capsules (produced by Sanjiangyuan) or dried raw Cordyceps bought direct from China, as some of them claimed that Cordyceps-based drugs they bought in Russia 'did not work.' Three people who agreed to talk to us explained that they used *Cordyceps sinensis* not as a substitute but in addition to the treatment offered by their doctors. Of the 50 pharmacies observed in Moscow in August 2023, five of them had Cordyceps-based products, but all of them were *Cordyceps militaris*.

Russian mycologist Mikhail Vishnevsky (b. 1973) (editor, businessman and the head of the Russian Association of Mycologists) calls himself 'a populizer of science' and 'a mycopharmacist' (Vishnevsky, 2024) and has authored several books on mushrooms. He saw an opportunity in 2014 to set up 'a first shop of artisan mushroom products,' such as butter with truffles from Crimea. He claims, on his website, that he has used *Cordyceps sinensis* to treat his own cognitive condition and received positive results ('felt 10–20 years younger'), and he has promoted knowledge about Cordyceps in Russia. His shop offers two types of Cordyceps: 'Russian cordyceps'—*Cordyceps militaris* powder capsules—and raw *Cordyceps sinensis* harvested from Nepal and India, which is not always available in the online shop. Interestingly, he states that his decision to supply *Cordyceps sinensis* from these parts of the Himalayas is due to 'better solarisation' of the Cordyceps habitats in these two countries, but it is more likely because foreigners are not allowed to harvest Cordyceps in China. He has warned his readers about counterfeit Cordyceps: 'Fake Cordyceps taste bitter' and '99% of the biologically active additives made with Cordyceps don't work' (Vishnevsky, 2024). For those who want an adventure and to taste authentic fungi, Vishnevsky has been offering tours priced at 16,000 USD per person to harvest *Cordyceps sinensis* in Nepal. The key attraction is the price for authentic *Cordyceps sinensis*—'3–4 times cheaper than the usual market price of 25–35,000 USD' (Fungiline, 2024).

Also, the Russian internet is sharing DIY recipes of how to grow *Cordyceps militaris*, and people are selling sample cultures (Warflop, 2020).

In parallel, a Russian Cordyceps industry has been emerging. The following pharmaceutical enterprises located all over Russia produce food products with Cordyceps. We also found at least 16 individual entrepreneurs producing products based on *Cordyceps militaris*: dried Cordyceps powder, concentrated tea and yoghurt, soap and pesticides, alongside the established pharmaceutical enterprises (Table 6.3).

To conclude this brief analysis, two aspects should be highlighted. First, foreign suppliers of natural *Cordyceps sinensis* in Bhutan, India, and Nepal do not present much competition to the Chinese suppliers, as over 90% of *Cordyceps sinensis* is harvested in China, and the issues they share are similar to those of Chinese collectors. Secondly, artificially grown, a cheaper *Cordyceps militaris* is a potential threat to *Cordyceps sinensis*. Relatively cheap cultivation enables companies from all over the world to enter the market. However, further survival in the market will require advanced technological facilities, smart marketing and attention to customers' needs and changing preferences.

Table 6.3 Russian Cordyceps-producing pharmaceutical companies

Enterprise	*Location*	*Products*
Pikodist	Moscow	Dried C. militaris
MZPOS Engelsk	St. Petersburg	Tea
NordPharm Ltd	Smolensk	Food additives
Alexnevskiy Express Ltd	Moscow	Dried Cordyceps
Galen pharmaceuticals	Barnaul	Food additives
Biolux Ltd.	St. Petersburg	Lozenges, Dried Cordyceps,
Biotica C Exporter	Moscow	Dried Cordyceps
Sashera Med	Biysk	Soap with C. militaris
Shishkin Ltd.	Novosibirsk	Lozenge
Biosynergia	Buryatia	Dried yoghurt with C. militaris
Zdorovaya Semiya	Biysk	Drink with C. militaris
ZEMBI	Moscow	Extracts of C. militaris
Milamed	Perm	Concentrated drinks
Sophphid	St. Peterburg	Extracts of C. militaris
Magnasvet	Krasnoarmeysk	Concentrated drink with C. militaris
SibPharmContract	Tomsk	Drink with C. militaris
RostGribTorg	Moscow	Dried C. militaris
ArtLife	Tomsk	Dried C. militaris
Samorazvitie	St. Petersburg	Food additive with C. militaris
Pharm Product	Barnaul	Tea
Altai Eco	Barnaul	Dried C. militaris

Note

1 In the East District, it is found in and around Yakla, Thegu, Serethang, Nathu La, Doklam, Bhutan La, and Lam Pokhari; in the North District, it is found in and around Zemu, Thay La, Kyongsey La, Yathang Lagyab, Samdong Lagyab, Toga Lagyab, Thumbyak, Thyapa, Goma Phu, Singhi Phyak, Dozem, Katao, Yumthang valley, Momey Samdong (commonly Yomey Samdong) valley, Lachung Thosa, Kishong valley, Singho Lake, and

the Paanch Pokhari area (Pradhan et al., 2020). In Uttarakhand, *Cordyceps sinensis* is harvested in Darma, Niti and Choudans valleys, Ralamdhura, Panchachuli base, Dharchula and Munsyari areas, Pindari catchment in Bageshwar district, Sutol, and Kanol in Chamoli district (Wallrapp et al., 2019; Negi et al., 2020).

References

Adhikari, J., Ojho, P., & Bhattari, B. (2016). Edible forest? Rethinking Nepal's forest governance in the era of food insecurity. *International Forestry Review*, *18*(3), 265–279 doi: 10.1505/146554816819501646

Bhakuni, H.S., & Chaunan, P.R. (2023). The Shauka community: The unsung tribe of the Central Himalayas in context to Rani Jasuli Devi. *IJFMR*, *5*(4), 4435.

Caplins, L., & Halvorson, S.J. (2017). Collecting Ophio*Cordyceps sinensis*: An emerging livelihood strategy in the Garhwal, Indian Himalaya. *Journal of Mountain Science*, 14(2), 390–402. doi: 10.1007/s11629-016-3892-8

Cannon, P.F., Hywel-Jones, N.L., Maczey, N., Norbu, L., Tshitila, Samdrup, T., & Lhundup, P (2009). Steps towards sustainable harvest of Ophio*Cordyceps sinensis* in Bhutan. *Biodiversity Conservation*, 18, 2263–2281.

Childs, G., & Choedup, N. (2014). Indigenous management strategies and socioeconomic impacts of Yartsa Gunbu (Ophio*Cordyceps sinensis*) harvesting in Nubri and Tsum, Nepal. *HIMALAYA 34*(1), 8–22.

Croizier, R.C. (1968). *Traditional medicine in modern China*. Harvard University Press: Cambridge.

Devkota, S. (2010). Ophio*Cordyceps sinensis* (Yarsagumba) from Nepal Himalayas: Status, threats and management strategies. In P.H. Zhang (Ed.), *Cordyceps resources and environment* (pp. 91–108). Grassland Supervision Center, Ministry of Agriculture, People's Republic of China: Xining.

Dong, C., Guo, S., Wang, W., & Liu, X. (2015). Cordyceps industry in China. *Mycology*, *6*(2), 121–129. doi: 10.1080/21501203.2015.1043967

Dorji, C. (2023, March 1). The magic fungus from the Himalaya. *Tashi Delek*. https://www.tashidelekmagazine.com/the-magic-fungus-of-the-himalayas/

Edwards, D.M. (1996). The trade in non-timber forest products from Nepal. *Mountain Research and Development*, *16*(4), 383–394.

Elkhateeb, W.A., & Daba, G.M. (2020). Review: The endless nutritional and pharmaceutical benefits of the Himalayan gold, Cordyceps; Current knowledge and prospective potentials. *Journal of Natural Product Biochemistry*, *18*(2), 70–77.

Filina-Kogan, V. (2022, June 7). Sent back by ten years: What's happening to cancer care in Russia. *News.Ru*. https://news.ru/society/otbrosili-na-desyat-let-chto-proishodit-s-onkologicheskoj-pomoshyu-v-rossii/?ysclid=lujgpe3emn673428949

Fungiline (2024). Mikhail Vishnevsky invites you to an expedition for wild Cordyceps in Nepal (May 18–29). www.fungiline.com/blog/mihail-vishnevskij-priglashaet-v-ekspediciziyu-za-dikorastushhim-kordiczepsom-v-nepal-18-29-maya/?ysclid=lujgzz2geb279545439

Heinen, J.T., & Yonzon, P.B. (1994). A review of conservation issues and programs in Nepal: From a single species focus toward biodiversity protection. *Mountain Research and Development*, *14*(1), 61. doi: 10.2307/3673738

Hooper, A. (2005, September 3). Chinese fungus poses eco-threat. *New Scientist*.

Huaon Research Institute (2021). *Forecast report on market competition and investment potential of Cordyceps in China (2022–2027)*. Huaon: Beijing.

James, D.W. (1955). Chinese medicine. *Lancet*, 265, 1068–1069.

Lednev, G.R., Kryukov, V.Yu., & Tshernyshev, S.E. (2007). First record of Cordyceps militaris Fries in West Siberia. *Euroasian Entomological Journal*, 3, 253–254.

Lhamo, P. (2022, April 21). Illegal Cordyceps collectors in Lunana. *Kuensel*.https://kuenselonline.com/illegal-cordyceps-collectors-in-lunana

Li, X., Liu, Q., Li, W., Li, Q., Qian, Z., Liu, X., & Dong, C. (2019). A breakthrough in the artificial cultivation of Chinese Cordyceps on a large-scale and its impact on science, the economy, and industry. *Critical Reviews in Biotechnology*, 39, 181–191.

Ministry of environmental resources of Buryatia (2023). Krasnaya Kniga. Rastenia Buryatii. Ulan Ude. burpriroda.ru

Moudgil, M. (2019, January 4). Cordyceps, the Himalayan Viagra kindles fortunes, and begets plunder. *Mongabay*. https://india.mongabay.com/2019/01/cordyceps-the-himalayan-viagra-kindles-fortunes-begets-plunder/

Mucelli, A. (2018). Relevance of Western medicine and TCM in the Chinese and European markets: An overview. In A. Mucelli & F. Spigarelli (Eds.), *Healthcare policies and systems in Europe and China: Comparisons and synergies*. World Scientific: Singapore.

Negi, C.S. (2017). The changing socio-economic profile of the Shaukas and Rangs vis-à-vis loss of agro-diversity, and yartsa gunbu: A case study of the Askote Conservation Landscape, District Pithoragarh, Kumaun Himalaya. In R. Chand, E. Nel, & S. Pelc (Eds.), *Societies, social inequalities and marginalization marginal regions in the 21st century*. Springer: Cham.

Negi, C.S., Koranga, P.R., & Ghinga, H.S. (2006). Yar tsa gumba (*Cordyceps sinensis*): A call for its sustainable exploitation. *International Journal of Sustainable Development and World Ecology*, 13, 1–8.

Negi, C.S., Joshi, P., & Bohra, S. (2015). Rapid vulnerability assessment of yartsa gunbu (Ophio*Cordyceps sinensis* [Berk.] G.H. Sung et al) in Pithoragarh District, Uttarakhand State, India. *Mountain Research and Development*, *35*(4), 382–391.

Negi, V.S., Rana, S.K, Giri, L., & Rawal, R.S. (2020). *Caterpillar fungus in the Himalaya: Current understanding and future possibilities*. G.B. Pant National Institute of Himalayan Environment, Kosi-Katarmal, Almora, Uttarakhand, India.

Palden, T. (2017, July 17). Cordyceps collectors prefer market to auctions. *Kuensel*. https://kuenselonline.com/cordycep-collectors-prefer-market-to-auctions/

Pant, B., Rai, R.K., Wallrapp, C., Ghate, R., Shrestha, U., & Ram, A. (2017). Horizontal integration of multiple institutions: Solutions for Yarshagumba related conflict in the Himalayan region of Nepal? *International Journal of the Commons*, 11, 464–486.

Paterson, R.R.M. (2008). Cordyceps—A traditional Chinese medicine and another fungal therapeutic biofactory? *Phytochemistry*, 69, 1469–1495.

Paudel, S. (2023, March 23). Cordyceps harvesters undeterred by lockdown ban in Nepal Himalayas. *Third Pole*. https//www.thethirdpole.net/en/nature/opinion-cordyceps-harvesters-undeterred-by-lockdown-ban-in-nepal-himalayas/

Pauls, T., & Franz, M. (2013). Trading in the dark—The Medicinal Plan Production Network in Uttarakhand. *Journal of Tropical Geography*, 34, 229–243. doi: 10.1111/sjtg.12026

Persistence Market Research (PMR). (2020). *Traditional Chinese medicines market global industry analysis 2015–2019 and forecast 2020–2030*. PMR: NY.

Polo, M. (1845). *The travels of Marco Polo*. Oliver & Boyd: Edinburgh.

Pradhan, B.K. (2016). Caterpillar mushroom, Ophio*Cordyceps sinensis* (Ascomycetes): A potential bioresource for commercialization in Sikkim Himalaya, India. *International Journal of Medicinal Mushrooms*, *18*(4), 337–346.

Pradhan, B.K., Sharma, G., Subba, B., Chettri, S., Chettri, A., Chettri, D.R., & Pradhan, A. (2020). Distribution, harvesting and trade of yartsa gunbu (Ophio*Cordyceps sinensis*) in the Sikkim Himalaya, India. *Mountain Research and Development*, *40*(2), 41–42.

Ren, W., Wang, C., Zhao, R., Li, H., Zhang, Q., & Cai, D. (2021). Artificial Cordyceps mycelium by submerged fermentation of Hirsutella sinensis HS 1201 using rice bran hydrolysate as substrate. *Environmental Quality Management*, *31*(1), 109–118.

Sharma, N. (2020). Challenges faced by the Bhotias for their livelihood and preservation of culture. *International Journal of Sociology and Anthropology*, 12920, 51–58.

Shrestha, U.B., & Bawa, K.S. (2013). Trade, harvest, and conservation of caterpillar fungus (Ophio*Cordyceps sinensis*) in the Himalayas. *Biological Conservation*, 159, 514–520.

Singh, S. (2016). *The local governance: Politics, decentralization and environment*. Oxford University Press: New Delhi.
Thapa, S. (2017, September 21). Bhutanmbe Cordyceps collectors eye return to yak herding. *Third Pole*. https://www.thethirdpole.net/en/climate/bhutans-cordyceps-collectors-eye-return-to-yak-herding
Timmerman, L., & Smith-Hall, C. (2020). Commercial medicinal plant collection is transforming high-altitude livelihoods in the Himalayas. *Mountain Research and Development 39*(3). doi: 10.1659/MRD-JOURNAL-D-18-00103.1
Tshetem, K. (2019). Sustainable Cordyceps harvesting in Bhutan: Interview with David Crow at Plant Medicine Summit 2019. https://s3.amazonaws.com/tsnshift/summits_transcripts/PlantMedicineSummit2019-0319-1100-KarmaTshetem-Transcript.pdf
Verified Market Research (VMR) (2023). *Global Cordyceps sinensis and militaris market size, status and forecast to 2030*. VMR: Washington.
Vishnevsky, M. (2024). personal website. https://michailvishnevsky.com/
Wallrapp, C., Keck, M., & Faust, H. (2019). Governing the yarshagumba 'gold rush': A comparative study of governance systems in the Kailash landscape in India and Nepal. *International Journal of the Commons*, *13*(1), 455–478.
Wangchuk, K., & Wangdi, J. (2015). Mountain pastoralism in transition: Consequences of legalizing Cordyceps collection on yak farming practices in Bhutan. *Pastoralism*, *5*(4). https://doi.org/10.1186/s13570-015-0025-x
Warflop (2020). Cordyceps militaris—Cultivation. *Pikabu*. https://pikabu.ru/story/cordyceps_militaris_kultivatsiya_7107878?ysclid=lujhjcbp4b275254658
Winkler, D.C. (2009). Fungus (Ophio*Cordyceps sinensis*) production and sustainability on the Tibetan Plateau and in the Himalayas. *Asian Medicine*, 5, 291–316.
Xu, Q., Bauer, R., Hendry, B.M., Fan, T.P., Zhao, Z., Duez, P., Simmonds, M.S.J., Witt, C.M., Lu, A., Robinson, N., Guo, D., & Hylands, P.J. (2013). The quest for modernisation of traditional Chinese medicine. *BMC Complementary and Alternative Medicine*, *13*(132).

Part III

Policies for Sustainable Management of *Cordyceps sinensis*

China has the largest pool of *Cordyceps sinensis* in the world, and thus it has become the country's duty to manage it sustainably while meeting the growing demand for rare medicinal fungus, ensuring that future generations will be able to observe the species in its natural habitat. The country has recently achieved the elimination of absolute poverty and has had an active role in the push for biodiversity conservation.

In the previous part, we discussed the complexity of the supply chain in the production of Cordyceps-based products. Following the traditional logic of the invisible hand of the market, the industry should be able to develop efficiency and self-regulation. However, two key vulnerable actors can be identified: primary suppliers and consumers, alongside the endangered species itself—Cordyceps. Thus, protecting vulnerable actors and introducing sustainable practices among business enterprises is a task for the national and regional authorities.

The logic of Part III is as follows. For centuries, the TCM industry was unregulated, and the laws regulating fungi-harvesting practices in highland pastures are still new. It took time to formulate them, and it takes time to implement and start applying them in practice (Chapter 7). Cordyceps consumers face health safety as an issue. A scientific enquiry into the level of arsenic content in *Cordyceps sinensis* in 2016 touched upon another complex issue. Over centuries, TCM herbs and fungi have proven to provide necessary remedies; however, no one researched what the maximum level of heavy metals in Cordyceps was acceptable for human health. The public debates over the arsenic content in *Cordyceps sinensis* have demonstrated the importance of the consistency of medical research and reliable information dissemination among businesses, traders and consumers of Cordyceps (Chapter 8).

Protecting primary fungi suppliers—rural collectors—has also meant revising the system of access to Cordyceps-producing pastures. Chapter 9 gives a historic review of the land reforms in China, which has been a gradual process of implementing new land rights, with implications for collectors and the whole supply chain of *Cordyceps sinensis*. Chapter 10 explains the impact of climate change on the potential survival of the Cordyceps species and discusses current conservation measures. Indeed, the rural collectors of Cordyceps are the most vulnerable

DOI: 10.4324/9781003429753-10

group. Many of them have utilized the practice of collecting Cordyceps as their main income strategy against the heavy background of losing profitability in their other activities, such as animal husbandry, as a result of climatic and environmental conditions, man-made depletion of natural resources, and partial resettlement. Chapter 11 discusses this vulnerability and the historic and current measures adapted in the Cordyceps-producing provinces to alleviate rural poverty.

7 Market Regulation of *Cordyceps sinensis*

Jiping Sheng, Yijing Xin, Wenfan Su, Xinyi Xu, and Lin Shen

Introduction

Cordyceps sinensis is found in China in the Qilian Mountains, northwestern Yunnan, Western Sichuan Plateau, and most parts of the Himalayas, including Tibet, Qinghai, Sichuan, and other provinces (Liang et al., 2022). Due to the special and fragile environment of *Cordyceps sinensis* and the limited amount produced naturally, the demand for medicinal materials is exceeding natural supply. At present, there is no officially accepted artificial cultivation technology of this species of Cordyceps, and market supply and demand are seriously unbalanced. Therefore, the market regulation of *Cordyceps sinensis* is particularly important. Qinghai Province and the Tibet Autonomous Region have respectively introduced provincial regulations for the management of *Cordyceps sinensis* (Zhang et al., 2021). Ganzi Prefecture of Sichuan Province has also issued regulations along with other relevant county-level administrative departments. At the same time, the Qinghai Province *Cordyceps sinensis* Association has issued their own industry standards. These regulations provide a basis for establishing practices for the management and marketing of *Cordyceps sinensis* resources.

Laws and Regulations

At the national Chinese level, resource management regulations include environmental protection laws and grassland laws. These national regulations provide a legal basis for the protection of *Cordyceps sinensis*. For example, according to Article 51, Water and Soil Conservation Law of the People's Republic of China, if anyone acts in violation of this law, by collecting black moss, or if they shovel turf, dig up tree stumps or excessively dig up *Cordyceps sinensis*, licorice, or ephedra in any key preservation area or key control area, the water administrative department of the people's government at or above the county level will confiscate the illegal gains and impose a fine of no less than the amount of but no more than 5 times the illegal gains upon the violator; or if there are no illegal gains, they may impose a fine of no more than 50,000 yuan upon the violator (Cui, 2016, p. 49).

DOI: 10.4324/9781003429753-11

The Regulations of the People's Republic of China on Grassland Law and the Regulations of the People's Republic of China on Wild Plant Protection regulate *Cordyceps sinensis* collection and maintain the natural environment of the grassland. The Tibet Autonomous Region has also formulated 'Interim Measures for the Collection and Management of *Cordyceps sinensis* in the Autonomous Region,' in order to safeguard the legitimate rights and interests of all parties and promote the healthy and orderly development of the Cordyceps trading market according to the Law of the People's Republic of China on Consumer Rights Protection.

According to the Tibet Autonomous Region Interim Measures, the prospective collector of Cordyceps should obtain a collection certificate. The collection certificate is issued to local people living within each Cordyceps production area. In addition, the agricultural administrative department of the Cordyceps production area scientifically and reasonably determine the boundaries of the Cordyceps collection area and the amount of Cordyceps to be collected. It is forbidden to collect Cordyceps without a license or collect Cordyceps in forbidden areas. Cordyceps collectors should protect the ecological environment and the animal husbandry construction facilities of grassland. They must not use pest control tools, which can be destructive to the grassland vegetation.

The administrative departments of agriculture and environmental protection in the Cordyceps production areas are responsible for strengthening law enforcement. Such departments should investigate and punish illegal harvesting of Cordyceps. If the collection certificate is not obtained or the Cordyceps is not collected in accordance with the collection certificate, the agricultural administrative department should order the collector to stop harvesting the Cordyceps and confiscate the illegally collected Cordyceps and the associated illegal income. Where the collection of Cordyceps causes damage to forests and forest land, the forestry administrative department imposes penalties in accordance with the provisions of the Forest Law of the People's Republic of China and similar relevant laws and regulations (Cheng et al., 2015).

According to the law, the acquisition of Cordyceps fungi should be licensed. Citizens or enterprises that purchase Cordyceps should hold a business license and a valid identification certificate issued by the administrative department for industry and commerce to process the trade of Cordyceps. Farmers and herdsmen who engage in the acquisition of Cordyceps should apply for an acquisition license from the county-level agricultural administrative department. If a pharmaceutical production and operation enterprise already has a Pharmaceutical Production License and a Pharmaceutical Business License, it does not need to apply for the Cordyceps acquisition license.

Cordyceps buyers should go to a government-designated trading point to acquire Cordyceps, according to the management regulations established by the local government. Individuals or enterprises that intend to sell Cordyceps should have a fixed business place as well as the aforementioned business license. However, farmers and herdsmen can sell Cordyceps that they collect themselves with just a Cordyceps acquisition license.

The logo and the physical object of the packaged Cordyceps must be original. And the distribution unit, address, origin, net content, year of collection, date of packaging and safe use instructions should be indicated in both Tibetan and Chinese characters. Packaging of Cordyceps should comply with the Measures for the Supervision and Administration of Quantitative Packaging Commodities Measurement.

For Cordyceps trading activities, the following acts are prohibited: repelling and controlling other people's normal trading activities; manipulating prices in order to set a monopoly price or monopolizing operations; the application of measuring instruments prohibited by the state, or measuring instruments that have not been verified; forgery of Cordyceps origin—using other factory names, sites and quality marks; posting false advertising in order to mislead and deceive consumers (Tibet Autonomous Region People's Government, 2009).

Market promoters, counter operators and trade fair organizers should strengthen the management of the operators of Cordyceps trading sites and supervise the quality of Cordyceps, sign quality agreements with the operators, clarify the responsibilities of both parties, and implement quality commitments. Market promoters, counter operators and trade fair organizers should be responsible for illegal dealers selling fake and inferior Cordyceps and bear joint and several liability (Tibet Autonomous Region People's Government, 2009).

The administrative department for industry and commerce carries the responsibility to supervise and inspect unlicensed operations, over-range operations, and transactions that disrupt market order and violations of laws and regulations in Cordyceps trading activities. The agricultural administrative department is responsible for supervising the acquisition of Cordyceps, in accordance with the law. The release and use of the Cordyceps acquisition license should also be checked by them, and the unlicensed acquisition of Cordyceps should be punished. The product quality supervision and management department supervise and inspect the quality of Cordyceps, as defined by the law. Public security institutions are responsible for investigating and punishing illegal and criminal activities that use Cordyceps trading to engage in fraud (Shi, 2022).

The relevant laws and regulations in Tibet require that a Cordyceps collection plan should be decided at county level according to local Cordyceps resources, and reasonably determine the collection area, the planned collection amount, an appropriate collection amount, the number of collection personnel, the collection period, and prohibited areas (Xiao, 2019). The regulations require Cordyceps collectors to protect the natural environment of the grasslands and surrounding construction facilities. The corresponding supervision and inspection system has been established. This method has had a positive effect on the sustainable collection of *Cordyceps sinensis* (Tenzin Phuntshogs, 2023).

In fact, the implementation of rules for the collection and management of *Cordyceps sinensis* at the county level is more detailed than the provincial regulations. For example, Qinghai Golog clearly stipulates that people from other areas should not be allowed to enter the Cordyceps production area to collect the precious fungi. Ganzi Prefecture of Sichuan Province has also issued an interim measure for the collection and management of *Cordyceps sinensis*, clarifying that the competent authority for the collection and management of *Cordyceps sinensis* is the state agricultural administrative department, which has incorporated *Cordyceps sinensis* into its resource management list (Cheng et al., 2015).

Local Standards for *Cordyceps sinensis*

Market regulation of *Cordyceps sinensis* not only covers collection and trade behavior but more importantly the standardization for Cordyceps. Qinghai Province and Tibet, the key *Cordyceps sinensis* production areas, have established local standards for Cordyceps.

Local Standard in Qinghai Province

Under the recommendation of the Qinghai Provincial Bureau of Quality and Technical Supervision, the Qinghai Provincial Standardization Association, the Qinghai Provincial Quality and Technical Supervision Information Management Center and the Qinghai Provincial Product Quality Inspection Institute and related processing and sales enterprises established a joint supervising team in 2007 in order to formulate local standards for *Cordyceps sinensis* (DB 63/670). This standardization aims to ensure the quality of *Cordyceps sinensis*, protect the interests of consumers, and also to protect Cordyceps habitats. This standard specifies the classification, technical requirements, test methods, packaging, storage and other aspects of processing *Cordyceps sinensis* (Zhang, 2010).

In order to meet the needs of production capacities and the market demand and protect the interests of both producers and distributors, as well as reduce the complexity of the complaint process, the standards have included the chapter 'Qualitative identification of Cordyceps' as a guide to the preliminary identification of Cordyceps (Zhang, 2010).

Cordyceps sinensis is sparse and expensive. Sometimes criminal sellers use flour and talcum powder to make non-worm wooden materials look like Cordyceps to deceive consumers and obtain illegal profits. In other instances, they sell 'pseudo Cordyceps,' which is Cordyceps that has not been formed via bat moth larvae (Zhang, 2010). The consumer cannot easily distinguish this type of Cordyceps. Inferior Cordyceps refers to artificial treatment—adding preservatives, using colorants, and adding metal substances to Cordyceps. These added substances can affect the look, weight or color of the product and deceive customers (ibid.).

The classification of *Cordyceps sinensis* has been chaotic in the market for a significant period of time. To stabilize the market, state experts began measuring the length and diameter of Cordyceps and then established the Cordyceps grading system in 2007 (Zhang, 2010). Subsequently, in 2018, the Chinese Society of Traditional Chinese Medicine promulgated the Commercial grades for Chinese Materia medica—CORDYCEPS (T/CACM 1021.33–2018)— which is applicable to the evaluation of commodity specification grade in the production, circulation and use of *Cordyceps sinensis* herbs (Chinese Association of Traditional Chinese Medicine, 2018).

Table 7.1 Classification of *Cordyceps sinensis*

Grade	*Commonalities*	*Differences*
First		<1,500 strips per kilogram, no broken grass, no threaded strips, no deflated grass, no dead grass, no black grass.
Second	This product consists of the body of a worm with fungal cysts growing from the head. It is like a silkworm, 3–5 cm long, 0.3–0.8 cm in diameter; the surface is dark yellow to yellowish brown with 20–30 rings, and the rings near the head are finer; the head is reddish brown; there are 8 pairs of feet, and 4 pairs of feet in the middle of the body are more obvious; the texture is brittle and easy to break, and the cross section is slightly flat and yellowish-white. Cotyledons elongated cylindrical, 4–7 cm long, about 0.3 cm in diameter; surface dark brown to brownish brown, with fine longitudinal wrinkles, slightly widened in the upper part; texture pliable, cross section white. Slightly fishy smell, slightly bitter taste.	1,500–2,000 strips per kilogram, no broken grass, no worn strips, no deflated grass, no dead grass, no black grass.
Third		2,000–2,500 strips per kilogram, no broken grass, no worn strips, no deflated grass, no dead grass, no black grass.
Fourth		2,500–3,000 strips per kilogram, no broken grass or worn strips.
Fifth		3,000–3,500 strips per kilogram, no broken grass, no threaded strips.
Sixth		3,500–4,000 strips per kilogram, no broken grass, no threaded strips.
Seventh		4,000–4,500 strips per kilogram, no broken grass, no threaded strips.

Source: Chinese Association of Traditional Chinese Medicine (2018).

Different from other standards, special provisions are made for *Cordyceps sinensis*. The purpose for such regulation is to protect the interests of consumers. The special regulations are as follows. Firstly, the product has to be pure and natural, collected in the wild. During the processing of the product, it must not be treated with chemicals (e.g., using sulfur to smoke it). Secondly, no preservatives should be added, and colorants should not be used. Third, metal materials or heavy metals should not be added. Fourth, any other matter that may affect the look, weight or color of the product is not allowed. Fifth, if the Cordyceps breaks, it must not be put back together.

Geographical Indication of Products from Nagqu, Tibet

In order to regulate the collection and trade behavior of Cordyceps, Tibet Nagqu authority has applied for geographical indication for Cordyceps from this region. This approach further strengthens the market regulation of Cordyceps on the basis of local standards.

A geographical indication product is one that is produced in a specific geographical area and whose quality, reputation or other characteristics are essentially dependent on the natural and human factors of the place of origin. This system has become part of China's intellectual property and plays an increasingly important role in improving the quality of featured products and promoting regional economic development and foreign trade (Gao, 2013).

In order to regulate the trading of geographical indication products in Nagqu, Tibet and maintain the quality of specialty products, since 2012, the General Administration of Quality Supervision, Inspection and Quarantine has accepted an application for the protection of landmarks for *Cordyceps sinensis* in Tibet. At the same time, the local standard titled 'Geographical Protection Products Tibet Nagqu Cordyceps' was also approved for release on March 28, 2012, and implemented on April 28, 2012.

In recent years, the price of *Cordyceps sinensis* has been rising. Nagqu produces some of the best quality, and therefore its price is much higher than that of average *Cordyceps sinensis* products. There are many traders who pretend their Cordyceps is from Nagqu to deceive consumers and make illegal profits. In order to prevent this, from 2011 to 2012, the Nagqu Quality Supervision Bureau established the Tibet Nagqu *Cordyceps sinensis* Geographical Indication Product Protection Leading Group, who designed a special geographical indication logo for Nagqu *Cordyceps sinensis* in Tibet and developed the network, telephone and SMS anti-counterfeiting query system. The Bureau also drafted 'Measures for the Administration of Geographical Indication Products of *Cordyceps sinensis* in Tibet.' These practices have led to a relatively complete management system, which effectively protects the interests of consumers and maintains the reputation of the Nagqu Cordyceps. The standard stipulates the basic requirements for the protection area, technical requirements, inspection methods, and the packaging and logos of Nagqu *Cordyceps sinensis* in Tibet.

The geographical indication product standard of Nagqu Cordyceps includes grading and sensory indicators to distinguish its quality.

Tibet Nagqu Cordyceps is a rare resource associated with a specific area and is highly priced. Harvesting fungi provides additional income for local farmers and herdsmen and is important for local economic development. The successful formulation of local standards for GI products has played an active role in protecting local brands, promoting local economic development, and regulating the local

Table 7.2 Sensory indicators

Item	*Index*
Shape	The body is 3 cm–5 cm long and has a diameter of 3 mm–8 mm. The surface is yellow to brown. The head is reddish brown, and the abdomen has 8 pairs of feet.
Smell and taste	Cordyceps tastes bitter and has a distinctive mushroom-like smell.
Impurity	No visible foreign objects
other	The appearance is complete and dry

Source: Qinghai Bureau of Quality and Technical Supervision (2010).

market. The successful implementation of local standards helps to accurately define the origin of *Cordyceps sinensis*, clarify its sensory characteristics, and determine its physical and chemical indicators. In the implementation process, local standards can effectively prevent the spread of fake *Cordyceps sinensis*. It also can reduce disputes about the quality of Cordyceps and maintain the reputation and branding of a Cordyceps product.

Regulation of Cordyceps Industry Associations

Industry associations are non-governmental organizations that play an important role in market regulation, disciplining industry behavior, coordinating peer-to-peer interests, maintaining fair competition and legitimating interests between industries, and promoting industry development.

Overview of the Qinghai Province **Cordyceps sinensis** *Association*

The Qinghai Province *Cordyceps sinensis* Association was established on September 19, 2007, registered by the Qinghai Provincial Civil Affairs Department. It is a public, non-profit group with independent legal status. It focuses on scientific research, production enterprises, individual businesses, self-employed workers, and brokers engaged in the *Cordyceps sinensis* industry (Lu, 2007). The association promotes 'civil management [and] public benefit' and the principles of 'self-education, self-management, and self-service.' Its purpose is to unite the province's Cordyceps operators to focus on the cultivation, collection, processing and trade of Cordyceps and jointly create the Qinghai Cordyceps brand. Its establishment was meant to further promote the development of the Cordyceps business in Qinghai (The Qinghai Province *Cordyceps sinensis* Association, 2007).

The association is closely focused on the sustainable development of the *Cordyceps sinensis* industry, to serve the government, society, enterprises, and the general public and improve the Qinghai Cordyceps brand (Guo, 2011). It is involved in the active expansion of the industry's self-discipline, the selection of industry representatives, regulation of industry services, and industry coordination, and serves as a bridge to connect different parties. The association promises that if the *Cordyceps sinensis* sold from the member units does not meet the standards, it will deal with the quality inspection and other relevant departments itself in the interests of the consumers (Tibet Autonomous Region People's Government, 2009).

The Cordyceps Association's Functions and Role

According to document No. 36 Guobanfa (2007) of the General Office of the State Council, 'Several Opinions on Accelerating the Reform and Development of Industry Associations and Chambers of Commerce' and the Qinghai Provincial Government General Office Qingbanfa (2007) No. 178 'On the Reform and Development of Industry Associations and Chambers of Commerce,' the association stipulates that 'Five Functions' will be implemented: disciplining of the industry, industry service and representation, industry coordination, and other authorized functions in the name of carrying out the following specific work:

(1) Bridge the gap between government and *Cordyceps sinensis* operators
The association serves as a bridge between *Cordyceps sinensis* operators and the government, accepts reasonable suggestions from operators and fulfils reasonable requirements; it provides legal and policy services as much as possible; organizes various *Cordyceps sinensis* exchange meetings and product promotion meetings; and solves problems concerning Cordyceps management.
(2) Legal promotion and environmental protection
The association actively publicizes laws, regulations and current standards related to *Cordyceps sinensis*, to enhance awareness of protecting the natural environment. It collaborates with governments at all levels in resource areas to stop illegal digging and excavation and create conditions for the sustainable development of the *Cordyceps sinensis* industry. At the same time, it actively cooperates with the quality inspection department to implement local standards for Qinghai *Cordyceps sinensis* and maintain its reputation. It has formulated and implemented resources on 'Classification of *Cordyceps sinensis* Fungal Germplasm,' '*Cordyceps sinensis* Sperm,' 'Qinghai Cordyceps Fungus Powder Products,' and 'Qinghai Natural Cordyceps Preservation Technical Regulations' to promote the sustainable development of the province's Cordyceps industry.
(3) Branding and market regulation of *Cordyceps sinensis*
The Association enhances the brand awareness of Qinghai *Cordyceps sinensis* by cleaning up and rectifying the market and conducting scientific inspections on the quality of products sold. Operators outside Qinghai and Tibet are issued market access permits to conduct legal operations at specialized business sites. The Association has cracked down on counterfeiting, to prevent inferior *Cordyceps sinensis* from entering the market. It is resolutely against anyone trying to pass off overseas *Cordyceps sinensis* as Tibetan *Cordyceps sinensis*, to ensure that the credibility of Tibetan Cordyceps is not infringed.
(4) Association member certification and capability enhancement
It issues plaques to its member units and the governing units of the *Cordyceps sinensis* Association to encourage consumers to purchase Cordyceps from listed units. Through evaluation and listing, operators' abilities and levels of 'independent operation, self-discipline, self-development, and self-improvement' are continuously improved.
(5) External promotion and trade expansion
It actively carries out external publicity and foreign trade promotion services. By publishing various briefings, materials, books, albums, newspapers, and periodicals, as well as establishing networks, applying for famous trademarks, and holding product exhibitions, the Association has increased the visibility of Qinghai *Cordyceps sinensis* and business units and individual operators to consumers. Through Qinghai *Cordyceps sinensis*, the Association members believe that they 'will make a breakthrough at the gate of the country, let

Qinghai understand the world and let the world know Qinghai' (The Qinghai Province *Cordyceps sinensis* Association, 2007).

(6) Market research and serving member units

Through in-depth investigation and research, the Association's experts analyze the industry and share their opinions and requirements with operators. They formulate service directions and measures, and provide members with market information and trends in the form of briefings etc., in a timely manner. They intend 'to work hard to do whatever the members ask for; take the initiative to do whatever the members need; and actively do whatever is beneficial to the members,' and to report 'the actual situation, do practical things, be pragmatic, seek practical results, and achieve results' (The Qinghai Province *Cordyceps sinensis* Association, 2007).

Cordyceps sinensis Trading Platforms

Cordyceps sinensis trading platforms are central to the broader Cordyceps market ecosystem. They are designed to facilitate the smooth exchange of this valuable fungus by providing a regulated environment that enhances transaction efficiency. By providing a centralized marketplace, a platform not only streamlines the buying and selling process but also fosters trust and transparency among its participants (e.g., http://www.dongchongxiac.cn/).

The Qinghai Jiuying Cordyceps International Trading Platform is China's largest online sales platform for Cordyceps (Cai, 2015). In order to ensure the quality of *Cordyceps sinensis*, the Qinghai Provincial Bureau of Quality and Technical Supervision has strictly implemented the local standards of Qinghai Province, which were formulated and issued by relevant departments. In order to allow consumers to purchase genuine *Cordyceps sinensis* from Qinghai, the Qinghai Jiuying Platform has also carried out sample testing to ensure the quality of natural and pure Cordyceps.

The platform's goal is to crack down on counterfeit and shoddy products and fraud; in Qinghai they often smash up such products. If fake Cordyceps enters the market, it will pose a serious threat to the health of prospective consumers. Therefore, the Cordyceps brand sold by Qinghai Jiuying Platform complies with various sanitary expectations and licensing conditions. In order to make the Cordyceps in the snow plateau better for consumers, the platform conducts random sampling for each batch of *Cordyceps sinensis* that enters the trading center to determine the grade and assess the quality. The test results are published and distributed to the network. These practices ensure the quality of *Cordyceps sinensis* traded and regulate the supply chain.

In addition to its online presence, the trade platform has a specialist sales center where customers can browse and purchase *Cordyceps sinensis* products in a traditional retail setting. This physical space provides an opportunity for customers to interact with experts and learn more about the benefits of *Cordyceps sinensis* to make informed choices for their purchases. The retail center is designed to provide a smooth offline experience with a wide range of products available for inspection and purchase, ensuring that customers have access to the highest quality *Cordyceps sinensis* regardless of their preferred shopping method. This is the first international *Cordyceps sinensis* trading center in China that integrates trading, testing, grade assessment, processing, tourism and service.

Trade and Export Regulations

Trade and export regulations for this precious commodity are strict due to *Cordyceps sinensis*' classification as an endangered species and its status as a medicinal product. Compliance with these regulations is critical to the sustainable trade and conservation of this invaluable resource.

As a cornerstone of Chinese herbal medicine, trade in *Cordyceps sinensis* in China must comply with the quarantine agreements, protocols, memoranda, and other provisions signed between the Chinese government and importing countries or regions. This means that products intended for export must not only comply with Chinese national laws and regulations but also meet the relevant legal and regulatory requirements of importing countries or regions.

In addition, *Cordyceps sinensis* has been listed as endangered under the Convention on International Trade in Endangered Species of Wild Fauna and Flora (CITES) to ensure its sustainable use and conservation. This convention requires member countries to impose strict controls on international trade in these species.

The 'Regulations on the Protection of Wild Plants of the People's Republic of China' requires customs authorities to carefully inspect and approve the movement of *Cordyceps sinensis* on the basis of permits issued by the national authority responsible for managing the import and export of endangered species (Central People's Government of the People's Republic of China, 2017). This regulation sends a strong message that illegal trade in wild plants will not be tolerated by the Chinese government, and those found guilty will face severe legal consequences.

In summary, the strict regulations on the trade and export of *Cordyceps sinensis* demonstrate China's commitment to the sustainable use and conservation of this medicinal fungus. Through such domestic legislation, adherence to international nature protection agreements, and enforcement of strict regulations, China aims to protect this invaluable natural resource for future generations.

References

Cai, P. (2015). Discussion on sustainable utilization of *Cordyceps sinensis* resource management in Tibetan area of Qinghai Province, in Proceedings of 2015 China Grassland Forum [in Chinese].

Central People's Government of the People's Republic of China (2017). Regulations on the Protection of Wild Plants of the People's Republic of China.

Cheng, Y., Qiu, Y., Peng, C., et al. (2015). Discussion on regulations of *Cordyceps sinensis* resource management [in Chinese]. *Shi Zhen National medicine and national medicine*, *26*(2), 449–450.

Chinese Association of Traditional Chinese Medicine (2018). Commercial grades for Chinese Materia medica—CORDYCEPS (T/CACM 1021.33–2018) [in Chinese].

Cui, G. (2016). Study on the legal system of grassland ecological protection in Nagqu region of Tibet [in Chinese]. Tibet University.

Gao, X. (2013). Review and prospect of *Cordyceps sinensis* local standard construction in Nagqu, Tibet [in Chinese]. *Management Informatization in China*, 1, 2.

Guo, Y. (2011). Brief analysis of development status and protection countermeasures of *Cordyceps sinensis* in Qinghai Province [in Chinese]. Qinghai Province Science and Technology, *18*(3), 39–42.

Liang, J., Li, X., Chen, J., et al. (2022). Effects of different storage temperature and time on egg hatching of Batmoth [in Chinese]. *Qinghai Journal of Animal Husbandry and Veterinary Medicine*, *52*(3), 13–17.

Lu, H. (2007, September 21). Qinghai *Cordyceps sinensis* Association was established [in Chinese]. *QingHai Daily*.

Qinghai Bureau of Quality and Technical Supervision (2010). DB63T 890-2010 Geographical Indication Products Qinghai *Cordyceps sinensis*.

Shi, X. (2022). The Ministry of Public Security announced the Kunlun 2022 special action food and drug field to supervise the handling of cases [in Chinese]. *China's Food Industry*, 10, 26–27.

Tenzin Phuntshogs (2023, August 8). Making income generation more secure and sustainable [in Chinese]. *Tibet Daily*.

The Qinghai Province *Cordyceps sinensis* Association (2007). About. http://www.qhdcxcxh.com/index.php/me/about

Tibet Autonomous Region People's Government (2009). Interim Measures for the Administration of Cordyceps Trading in the Tibet Autonomous Region [in Chinese].

Xiao, J. (2019). Rules on collection, management and legislation of *Cordyceps sinensis* [in Chinese]. *Edible Fungi of China*, *38*(4), 8–10.

Zhang, C., Tian, Z., Fan, Q., et al. (2021). Analysis on development status and sustainable utilization of *Cordyceps sinensis* resources [in Chinese]. *Edible Fungi of China*, *40*(10), 79–88.

Zhang, S. (2010). Interpretation of local standards for *Cordyceps sinensis*, China Standardization [in China]. *Standardization in China*, 11, 51–53.

8 Cordyceps Quality Control Management in the Framework of the Chinese Food Safety Policy

Jiping Sheng, Wenfan Su, Xinyi Xu, Wenjie Long, and Lin Shen

Background

Arsenic trioxide was used in Traditional Chinese Medicine (TCM) for centuries, but then it was banned and then rediscovered through Western science as anti-leukaemic and anti-parasite treatment (Au, 2011). Yet it is highly toxic. In TCM herbs, arsenic comes from the natural environment, for example from soil.

The discussion on what level of arsenic content is acceptable in China began in 2016. On February 4, 2016, the China Food and Drug Administration (CFDA) issued a reminder regarding the consumption of *Cordyceps sinensis* products, stating that monitoring and inspection had revealed arsenic content ranging from 4.4–9.9 mg/kg (Department, 2019) in tested products. CFDA advised that Cordyceps should not be used as either a food product or a medicine. Experts stated that long-term consumption of Cordyceps-based products could lead to arsenic overdosing and create associated health risks (Liu et al., 2018; Shao et al., 2020), considering the national safety standards for health foods specify an arsenic limit of 1.0 mg/kg (Khan et al., 2008; Sanders et al., 2015; Shen, 2018).

On February 27, 2016, the Food and Drug Administration of the Tibet Autonomous Region issued a statement affirming that *Cordyceps sinensis* is a traditional Chinese medicinal material listed in the *Chinese Pharmacopoeia* without a specified arsenic limit. The statement emphasized the importance of using *Cordyceps sinensis* under the guidance of a doctor (Cao et al., 2015).

The existing scientific research has not reached a consensus about the matter, particularly in light of the fact that unknown arsenic speciation was found in *Cordyceps sinensis* (Liu et al., 2018). There are also different standards for safety assessment. The WHO, FDA, and Hong Kong, China's arsenic content calculation standards were based on the maximum daily (weekly) intake (including the intake from other foods or drugs), while the standards of Canada, Singapore, Thailand, and South Korea were lower, to varying degrees, than the requirements of China's green industry standards for medicinal plants and preparations. The standards of the European Union were the most stringent. Even with the most lenient standards, there were still seven batches of medicinal materials from Qinghai that did not meet the requirements (Yuan, 2016).

DOI: 10.4324/9781003429753-12

The CFDA's consumption reminder sparked widespread media coverage and initiated public discussions on the safety of Cordyceps products (Xiao et al., 2012; Liu et al., 2016; Xiao et al., 2021). Some media publications claimed the low toxicity and safety of Cordyceps-based products (Yuan, 2016). At the same time, one month later, the pilot research work on Cordyceps health products was canceled, leading to significant financial repercussions for companies who were specializing in producing Cordyceps products such as the Very Grass range or had dried *Cordyceps sinensis* as its main product. These companies faced bankruptcy, as their main revenue source was forced off the market (Chen et al., 2013; Cui, 2019).

To assess the impact of the arsenic debate on collectors, traders, companies, and consumers in the Cordyceps industry, a group of us from the School of Agricultural Economics and Rural Development, Renmin University of China, conducted investigations in Xining City, Qinghai Province. We examined the impact on sales through interviews with local middlemen in Tibet-Qinghai and the Cordyceps-producing enterprises, and consumer behavior was assessed through online questionnaires.

Impact of the Arsenic Debate on Middlemen (Traders)

The arsenic public debate and standardization led to serious changes within the *Cordyceps sinensis* industry (Giri et al., 2016; Maghakyan et al., 2017; Jiang et al., 2018). The price of *Cordyceps sinensis* temporarily fell quite significantly. However, we found that after the release of the *Cordyceps sinensis* consumption recommendations, the price went back to normal or was not much affected.

Cross-cutting Analysis of Local Cordyceps Stores' Performance

We examined the impact of the debate on stores' performance between February and April 2016 through a comparison of sales in Cordyceps during the same period in 2015 (Table 8.1). The analysis focuses on the sales performance of shopkeepers and correlates it with various demographic characteristics.

Among male shopkeepers, 56.1% reported sales levels consistent with previous years, while 12.2% of shops experienced an increase; 7.4% witnessed a sharp decrease, and 14.9% observed just a slight decrease in sales. Additionally, 9.5% of male shopkeepers reported a significant increase in sales during this period. For female shopkeepers, 22.2% reported a sales level consistent with previous years, while a substantial proportion (55.6%) indicated no significant increase in sales during this period.

Examining different age groups, the majority of shopkeepers reported a sales level consistent with previous years, ranging from 50.0% to 61.9%. However, variations were observed within each age group. The age group 40–49 had the highest proportion of shopkeepers experiencing a slight decrease (19.6%) and a significant increase (24.3%) in sales. In contrast, the age group 50–59 had the

Table 8.1 Cross-analysis of sales from February to April 2016 compared to the same period in 2015

	Sales from February to April 2016 compared to the same period in 2015					*Chi-squared test (χ^2)*	*df*	*p*
	Decrease Sharply	*Decrease Slightly*	*Same as previous years*	*Increase Slightly*	*Increase Sharply*			
Gender						13.785	4	0.008
Male	7.4%	14.9%	56.1%	12.2%	9.5%			
Female	11.1%	11.1%	22.2%	55.6%	0.0%			
Age						15.897	16	0.460
18–29	8.1%	8.1%	51.4%	16.2%	16.2%			
30–39	6.7%	15.6%	55.6%	15.6%	6.7%			
40–49	6.5%	19.6%	52.2%	17.4%	24.3%			
50–59	14.3%	19.0%	61.9%	0.0%	4.8%			
Over 60	0.0%	0.0%	50.0%	25.0%	25.0%			
Academic qualifications						110.981	20	0.000
Did not attend school	33.3%	26.7%	33.3%	6.7%	0.0%			
Primary and below	11.8%	31.4%	56.9%	0.0%	0.0%			
Junior Middle School	3.1%	9.4%	65.6%	12.5%	9.4%			
High School or Technical Secondary School	0.0%	0.0%	65.8%	26.3%	7.9%			
College or Undergraduate	0.0%	0.0%	27.8%	44.4%	27.8%			
Postgraduate and above	0.0%	0.0%	0.0%	0.0%	100.0%			
Ethnicity						45.808	12	0.000
Hui people	9.0%	15.8%	58.6%	12.0%	4.5%			
Han people	0.0%	0.0%	18.8%	37.5%	43.8%			
Tibetan	0.0%	14.3%	57.1%	14.3%	14.3%			
Others	0.0%	100.0%	0.0%	0.0%	0.0%			
Type of shop						12.706	8	0.122

(*Continued*)

Table 8.1 (Continued)

	Sales from February to April 2016 compared to the same period in 2015					*Chi-squared test (χ2)*	*df*	*p*
	Decrease Sharply	*Decrease Slightly*	*Same as previous years*	*Increase Slightly*	*Increase Sharply*			
Individual	4.8%	12.5%	60.6%	13.5%	8.7%			
Company	12.0%	18.0%	44.0%	18.0%	8.0%			
Non-registered family stores	33.3%	33.3%	0.0%	0.0%	33.3%			
Type of shop investment						5.436	4	0.245
Sole Proprietorship	5.7%	14.3%	51.4%	18.1%	10.5%			
Joint Ventures	11.5%	15.4%	59.6%	7.7%	5.8%			

highest proportion of shopkeepers reporting a sharp decrease in sales (14.3%). Notably, no shopkeepers in the age group over 60 reported a sharp decrease or increase in sales during this period.

Regarding their academic qualifications, shopkeepers with a lower educational level experienced a decline in their sales. Shopkeepers without any schooling had the highest proportion of sharp decreases (33.3%) and slight decreases (26.7%) in sales. Similarly, shopkeepers with primary school education exhibited a significant proportion of slight decreases (31.4%) in sales. Conversely, shopkeepers with higher academic qualifications, such as college or undergraduate degrees, reported a higher proportion of sales consistent with previous years (27.8%) and a sales volume increase (44.4%, and sharp increase 27.8%). Notably, shopkeepers with postgraduate education or above reported a 100% increase in sales during this period.

Furthermore, the statistical analysis reveals a correlation between different ethnic groups and store sales in the three months following the arsenic debate. However, from a quantitative perspective, most individual businesses and companies observed sales that were not significantly different from previous years. Similarly, there was no statistical difference between different shop types and store sales in the three months after the debate, as indicated by the correlation coefficient.

Analysis of Factors Influencing Sales of **Cordyceps sinensis**

Sales of *Cordyceps sinensis* are vital for traders in the producing area, particularly Hui middlemen, whose family structure involves full-time housewives and men working outside the home. Several factors influence sales of *Cordyceps sinensis*. Firstly, individuals' physical health, as it is a medicinal material, and secondly, external economic development and national policies, such as government regulations, as it is considered a luxury product. Additionally, the sales performance of Hui traders is influenced by personal characteristics (such as age and education), social resources (such as partnerships and regular customers), and the overall economic level.

To investigate the impact of various factors on sales of *Cordyceps sinensis*, our research group conducted a questionnaire survey and identified independent variables including traders' personal characteristics (age, ethnicity, education), characteristics of their shop (type, size), sales (channels, customer base), and credit availability. The dependent variable was the sales volume of middlemen from February to April 2016. An ordinary least squares (OLS) regression model was used to analyze the factors influencing sales. The status of sales of the Cordyceps shop is expressed by the formula (1) as follows:

$$Y = \beta_1 + \beta_2 X_2 + \cdots \ldots + \beta_2 X_2 + \mu \tag{1}$$

Xi represents the various control variables that affect Cordyceps sales. The control variables here include traders' personal characteristics (age, ethnic education, shop operating years), shop condition (shop types, business years, shop capital forms), sales situation (sales channels, fixed number of customers), etc. Details of the variables are shown in Table 8.2.

Table 8.2 Variable definitions and assignments

Variable	*Definition*	*Variable assignment*
y	Sales	Large decrease = 1, small decrease = 2, similar to previous years = 3, small increase = 4, large increase = 5
X1	Age	Actual value
X2	Ethnicity	Minorities = 1, Han = 0
X3	Educational level	Did not attend school = 1, elementary school = 2, junior high school = 3, high school = 4, college or undergraduate = 5, graduate student and above = 6
X4	Type of shop	Individual business = 1, business = 0
X5	Operation time	Actual value
X6	Shop capital form	Sole proprietorship = 1, joint venture = 0
X7	Marketing channels	Face-to-face sales = 1, Network + Face-to-face sales = 0
X8	Fixed number of clients	Actual value
X9	Credit or not	Yes = 1, No = 0

Model Results and Analysis

In this paper, STATA 12 is used to estimate the OLS regression model of the sample data. The final model estimated results are shown in Table 8.3. The results indicate that the sales of traders are significantly correlated with their education level, years in business, and the number of loyal customers. However, there was no significant correlation between sales and the types of shops among ethnic groups, the form of shop funds, sales channels, or credit availability. The specific findings are explained as follows.

We examined the effect of personal characteristics. Education level has a positive and statistically significant impact on Cordyceps sales. Highly educated middlemen demonstrated better social skills and efficient management thinking. Additionally, educated middlemen are perceived by buyers as more capable of ensuring product quality and effective communication. However, local middlemen often have lower levels of education, leading to barriers in active socializing and in expanding customer numbers.

Age has a positive effect on sales. Older middlemen tend to have more established and successful businesses. Longer business tenure has positively affected sales, as accumulated experience contributes to increased sales volume. The negative impact of ethnicity may be attributed to communication difficulties in Mandarin and cultural differences while dealing with customers. Generally, Han middlemen have higher levels of education compared to their counterparts from other ethnic minorities

Self-employed middlemen were performing less successfully than those operating under a company structure. This can be explained by the perceived reputation and quality assurance associated with company-owned shops.

Table 8.3 Model results

Influence Factors	*Coefficient*	*Standard Error*
Age	0.035	0.006
Ethnic	-0.210	0.220
Educational attainment	0.235**	0.081
Type of shop	0.213*	0.049
Time of operation of the shop	-0.040	0.121
Shop capital form	-0.043	0.123
Marketing channels	0.253	0.153
Fixed number of clients	0.152***	0.048
Credit or not	0.258**	0.078
Constant term	-1.128*	0.529
Sample Quantity	157	
Virtual R2	0.404	

The legal status of shop funds also affects performance in sales. Joint venture middlemen performed better than sole proprietors, potentially due to having more resources and a stronger network of relationships associated with joint ventures.

The number of loyal customers has had a significant positive impact on sales. More permanent customers have provided a guaranteed sales volume.

Shops that solely conduct face-to-face sales outperformed those that sell both online and face-to-face. This finding is contrary to common sense but may be attributed to the prevalence of fake and expensive products in the online Cordyceps market.

Impact of the Arsenic Debate on Traders' Operations

We also conducted in-depth interviews with six typical Cordyceps trade middlemen to understand the specific conditions of resource acquisition and sales to assess the effect of the arsenic debate on their operations. Four typical Cordyceps shops were selected for analysis.

Impact on Specific Cordyceps Shops

In general, stores were impacted differently by the debate. While some stores remained largely unaffected due to their customer base and business partnerships, others faced challenges such as decreased sales and reduced profitability. Increasing competition and rising costs were additional factors affecting the profitability of the stores.

> *Store A* significantly increased the price of *Cordyceps sinensis*, resulting in substantial profits despite public concerns about arsenic contamination. The store used the *Cordyceps sinensis* Association of Qinghai Province's formula

to reassure customers that the arsenic content in their Cordyceps unlikely exceeds the local standards. Long-term consumers were not deterred by the new arsenic recommendation, but casual consumers showed some hesitation. To address anxiety about potential risks from Cordyceps, a strategy to educate casual customers about the safe daily intake of Cordyceps was implemented, and partnerships with loyal established customers were strengthened, which had a positive result. Sales remained consistent with previous years. Overall, the impact of arsenic contamination on Store A was minimal due to its customers and good partnerships with pharmacies.

Store B dismissed the notion of arsenic contamination but experienced a slight decrease in sales compared to previous years. Due to increased competition, the store lost a regular customer who had purchased a significant amount of *Cordyceps sinensis* in the past, and this contributed to the decline in sales. New customers were not effectively retained. Dismissal without factual explanation of the arsenic debate to customers contributed to this decline. The store also implemented price-reduction promotions but then faced increased prices in July, leading to regrets regarding the impact of the arsenic issue on sales.

Store C primarily engaged in wholesale business and remained largely unaffected by the arsenic debate. Shopkeeper C, with a shorter business duration, during the heat of the debate decided to focus on the quality of his products. Academic qualifications and broader horizons, along with cooperation from a younger brother who brought new customers from Guangzhou, a city in the rich Guangdong Province, contributed to the expansion of the business. There was an increase in sales as the store gained new customers spread across Hong Kong, Macao, and Taiwan. The overall impact on Store C's operations was minimal.

Store D was transparent about the arsenic contamination debate with their customers but still experienced some negative effects. Several casual visitors returned their purchases, and price reductions by pharmacies led to lower profits. The owner noticed the difficulty of operating the business and rising costs.

Shopkeeper D had first-hand knowledge of the arsenic debate due to having access rights over a *Cordyceps sinensis* pasture and was not dismissing it. However, their new customer development strategy was passive, relying heavily on their reputation to retain old customers. Dependence on old customers soon presented a challenge, and they experienced a profit margin decrease due to these customers switching to lower-priced alternatives.

In conclusion, rising operating costs and changing customer dynamics meant owners adopted a pessimistic outlook on the industry, although procurement needs from pharmacy enterprises still kept stores in business. Ultimately, strengthening relationships with existing customers and adopting a proactive marketing approach were the strategies for increasing the sales volume, despite customers' concerns raised by the arsenic contamination debate.

Impact of the Arsenic Debate on Cordyceps-Producing Enterprises

While we found that the price of *Cordyceps sinensis* remained stable, there were significant market price fluctuations. The impact of the new arsenic standards on Cordyceps-producing enterprises and the potential spillover effects were explored through interviews with three selected companies in Qinghai Province in March 2017.

Enterprise 1: Qinghai Spring Medicinal Resources Technology Co., Ltd is a Cordyceps-processing company, and their main product has been a pure *Cordyceps sinensis* powder. It is a Shanghai-Shenzhen board listed enterprise.

Enterprise 2: Qinghai Yushu Sanjiangyuan Cordyceps Technology Co., Ltd sells *Cordyceps sinensis* and is a three-board listed enterprise.

Enterprise 3: Qinghai Everest Cordyceps Herbal Medicine Co., Ltd is the largest processing company of *Cordyceps sinensis* artificial mycelium in Qinghai Province. Its main product is *Cordyceps sinensis* powder.

There are three main types of Cordyceps products: raw grass, processed products, and artificial mycelium products.

Qinghai Spring

On March 31, 2016, the Qinghai Food and Drug Administration issued a notice discontinuing the production of Cordyceps pure powder tablets, which accounted for 80% of the company's revenue. This suspension severely affected the company's operations and halted its rapid development. Its pharmaceutical business revenue dropped by 57.22% year-on-year, with Cordyceps pure powder revenue falling by 63.28%. Regional revenues also declined significantly. Despite introducing a different product called Yuancao, the sales situation, especially on e-commerce platforms, was not optimistic. The arsenic debate had a significant impact on Qinghai Spring's processing.

It took until 2017 for its Cordyceps business to recover. The company's flagship 'yuan grass' also sold poorly, and its main source of income changed from caterpillar fungus to advertising revenue.

Sanjiangyuan

The company experienced a minor decline in Cordyceps revenue (only 2%) from February to April 2016 due to the new arsenic standards. The development of the Sanjiang Yuanshencao tablet was affected due to the suspension of pilot work and research funding by the national health authority. The decline in Cordyceps revenue was primarily attributed to the overall economic slowdown and government regulations on corruption. While some new customers had concerns about Cordyceps' arsenic content, the company addressed these concerns through 24-hour customer service and the promotion of normal consumption standards. The company's wide-ranging advertising efforts aimed to enhance brand recognition and mitigate the impact on raw grass sales.

The reasons why Sanjiangyuan Cordyceps Technology Co., Ltd was not much affected by the debates on the excessive arsenic content of *Cordyceps sinensis* may be:

- The company has a good reputation and high consumer trust;
- mostly wealthy people use *Cordyceps sinensis* for health care;
- the period of consumption is usually limited, so most consumers will not exceed the intake for arsenic. Also, people who use *Cordyceps sinensis* for treatment purposes are mostly tumor patients; in many cases they do not have an alternative to *Cordyceps sinensis*.

Everest Company

According to Ms. Wangyuhua, head of the group marketing department, a study made by the company showed that the arsenic debate had minimal to no impact on their sales and revenues. Their *Cordyceps sinensis* products, such as *Cordyceps sinensis* powder, are artificially extracted and do not contain worm bodies. The product packaging clearly indicates the artificial production process, reassuring consumers of the safety of the product. Their main products, 100 Ling tablets and 100 Ling capsules, were considered drugs rather than ordinary food or health products. Typically, doctors purchased these drugs, and that's why consumers did not have concerns about arsenic contamination in this product. The crisis of the industry, initiated by the debate on the arsenic limit, did not have a spillover effect on other products. The corporate impact on artificial plant extracts was not noticeable

Impact of the Arsenic Debate on Consumer Behavior

The effect on consumer behavior was surveyed via online questionnaires. A total of 310 questionnaires were collected, of which 303 were valid. The survey subjects were located in Beijing, Shanghai, Guangdong, Shandong, Hunan, Jiangsu, Zhejiang, Tianjin, Henan, Anhui, Jilin, Fujian, and 12 other provinces and cities. In addition, there are also a few foreign consumers.

Study on Consumers' Knowledge of the Arsenic Contamination Debate

The respondents' knowledge of *Cordyceps sinensis* is shown in Table 8.4; 73.3 % of them had not heard of the consumption recommendation for *Cordyceps sinensis* products, issued by CFDA, and only 5.3% knew about its specific content. The recommendation had addressed consumers' poor knowledge of Cordyceps and was aimed at preventing potential health risks. Nearly 87.8% did not know that the arsenic content of caterpillar fungus exceeds the national standard.

Table 8.4 Knowledge about arsenic levels in Cordyceps

Index	*Project*	*Frequency*	*Percentage (%)*
Knowledge about the issued recommendation on *Cordyceps sinensis* products	Understand, know the details.	16	5.3
	Yes, but I don't know what it is.	65	21.5
	Never heard of it.	222	73.3
Is there a reason for the 'Consumer Guidelines on *Cordyceps sinensis*'?	YES	25	8.3
	NO	278	91.7
Do you know that long-term consumption of *Cordyceps sinensis* may lead to overdosing on arsenic?	YES	37	12.2
	NO	266	87.8

Cross-analysis of Levels of Consumer Awareness of the Arsenic Standard

In Table 8.5, a small percentage (19.1%) of men were aware that long-term use of Cordyceps may lead to arsenic overdose, while the majority (80.9%) were not aware. Only 8% of women knew of this potential correlation. There was a statistically significant difference between men and women in their understanding of the arsenic contamination.

The age group of 60 and above had the highest level of knowledge regarding arsenic levels in Cordyceps, while other age groups had similar levels of knowledge. There was no statistical difference in understanding between different age groups. Individuals with a high school education or below had the best knowledge about arsenic in Cordyceps. There was no statistical difference in understanding between different education groups.

People with chronic diseases had the highest level of understanding, followed by sub-healthy individuals, and then healthy individuals. However, there was no statistical difference in understanding between different health conditions. Individuals with a high annual net income had relatively good knowledge of arsenic in Cordyceps. However, there was no statistical difference in understanding between different income groups. The presence of elderly and non-elderly individuals in the family was associated with a better understanding of arsenic in Cordyceps.

Cross-analysis of Consumer Knowledge of Risks of Long-Term Consumption

From *Table 8.6,* we can see that:

(1) 27.1% of people who have ever purchased raw Cordyceps know that long-term consumption may mean they ingest more than the acceptable arsenic level, but only 8.6% of people who have never purchased the raw

product know the risks. Those who have ever bought Cordyceps have a better understanding of the arsenic debate.

(2) 20.5 % of people who have purchased *Cordyceps sinensis* processed products know that long-term use could lead to an overdose, while only 11.0% of people who have never purchased processed products know this. People who have consumed processed products have better knowledge about overdosing on arsenic.

(3) 38.9% of people who have taken *Cordyceps sinensis* over a long period know that long-term consumption may lead to overdose, while only 10.5% of people who have never taken it for a long period know this. Based on the correlation coefficient, the long-term use of *Cordyceps sinensis* has a statistical correlation with knowledge of arsenic content.

Table 8.5 Cross-analysis of knowledge of potential arsenic overdose and demographic characteristics

Main variables		*Do you know that chronic consumption of* Cordyceps sinensis *and its pure powder tablets may cause arsenic overdose? (%)*		*Sample share (%)*
		YES	*NO*	
Gender	Male	19.1	80.9	38.0
	Female	8.0	92.0	62.0
Age	20–29	10.3	89.7	51.2
	30–39	14.6	85.4	13.5
	40–49	11.9	88.1	13.9
	50–59	12.2	87.8	16.2
	60 and above	25.0	75.0	5.3
Academic Qualifications	High school and below	21.1	78.9	12.5
	College or undergraduate	11.5	88.5	63.1
	Postgraduate and above	9.5	90.5	24.4
Health Status	health	11.9	88.1	71.9
	Sub-health	12.3	87.7	26.7
	Disease	25.0	75.0	1.3
Net Annual Personal Income	\$50,000 and below	12.8	87.2	49.2
	\$50,000– \$80,000	11.6	88.4	14.2
	\$80,000–\$120,000	8.8	91.2	18.8
	\$120,000 and above	14.8	85.2	17.8
Is there an old person in the family?	YES	12.2	87.8	92.1
	NO	12.5	87.5	7.9

Annotations: * $p<0.05$; ** $p<0.01$; *** $p<0.001$

Table 8.6 Cross-analysis of knowledge of risk of chronic consumption and buying habits

Main variables		*Do you know that chronic consumption of* Cordyceps sinensis *and its pure powder tablets may lead to arsenic overdose (%)*		*Sample share (%)*
		YES	*NO*	
Experience with buying raw grass	YES	27.1	72.9	19.5
	NO	8.6	91.4	80.5
Experience with the purchase of processed products	YES	20.5	79.5	12.9
	NO	11.0	89.0	87.1
Long-term consumption of Cordyceps	YES	38.9	61.1	5.9
	NO	10.5	89.5	94.1

Study on Consumers' Willingness to Buy Cordyceps sinensis

Since many among those surveyed in our study did not know about the potentially dangerous arsenic content of the caterpillar fungus, we also tested consumers' willingness to buy before and after reading the national recommendation on Cordyceps consumption issued by the CFDA. We asked these questions:

Question 1: Is Cordyceps your first choice health product?
Question 2: Is Cordyceps still your first choice now you know that long term use may mean you ingest more than the acceptable level of arsenic?

We found that the number of people who would still buy Cordyceps as their first choice health product after reading the recommendation dropped from 72 to 27, a decrease of 62.5 %.

Question 3: If a friend wanted to buy health products, would you recommend Cordyceps?
Question 4: Would you still recommend Cordyceps to a friend having read the recommendation?

We can see that the number of people who would still recommend *Cordyceps sinensis* to friends dropped from 85 to 26, a decrease of 84.7 %.

Question 5: Will you buy Cordyceps even if the price is high?
Question 6: Now you know that long-term use may mean you exceed the acceptable intake of arsenic, will you still buy *Cordyceps sinensis*?
Question 7: If the answer to Question 6 is no, would you still buy *Cordyceps sinensis* if it was cheaper?

The number of people who thought that they would still buy *Cordyceps sinensis* even at a high price fell from 61 to 20, a decrease of 67.2 %. But it was clear that if the price of *Cordyceps sinensis* was much cheaper, people would still buy it.

Cross-analysis of Consumers' Intention to Buy Cordyceps

From *Table 8.7*, the following preliminary conclusions can be drawn:

1) 37.8% of men who have purchased *Cordyceps sinensis* said they would still consider buying it after learning about the arsenic content and the recommendation, while only 26.2% of women would still consider buying it. Although there are differences, the differences are not significant. From the correlation coefficient, there is no statistical difference in the willingness to purchase *Cordyceps sinensis* after learning about arsenic content.
2) People aged over 60 and between 40 and 49 were not willing to purchase the product still. Only 23.1% aged between 50 and 59 said they would still buy it. The other relatively young age groups are 40% to 50% willing to buy. Judging from the correlation coefficient, age has a statistical correlation to the desire to purchase *Cordyceps sinensis* even with knowledge of arsenic levels.
3) People with high school qualifications or below had the strongest willingness to still purchase *Cordyceps sinensis*. The higher the degree of education, the less people were willing to purchase *Cordyceps sinensis* after learning about arsenic levels. There is no statistical difference in purchase intentions of people with different education levels after learning about arsenic content.
4) The proportion of people with chronic diseases who were still willing to buy *Cordyceps sinensis* was 50%. Again, there was no statistical difference in purchase intentions of people with different health conditions after learning about arsenic levels.
5) According to the correlation coefficient, there was no statistical difference in the willingness of different income groups to purchase *Cordyceps sinensis* after learning about arsenic levels.
6) The willingness of people to still buy Cordyceps with an elderly person in the family compared to those with no elderly person in the family was relatively close. Again there was no statistical difference in the willingness of the two groups to purchase *Cordyceps sinensis* after learning about arsenic levels.

Cross-analysis of Consumer Awareness and the Long-Term Consumption of **Cordyceps sinensis**

The consumer survey conducted by our task group encompassed various regions across the country and primarily involved young and middle-aged consumers with

Table 8.7 Cross-analysis of purchase intentions after learning about arsenic levels in Cordyceps and demographic characteristics

Main variables		*Will you buy Cordyceps after learning that the arsenic content exceeds recommended levels? (%)*		*Sample share (%)*
		YES	*NO*	
Gender	Male	37.8	62.2	46.8
	Female	26.2	73.8	53.2
Age	20–29	47.1	52.9	43.0
	30–39	40.0	60.0	19.0
	40–49	0	100.0	13.9
	50–59	23.1	76.9	16.5
	60 and above	0	100.0	7.6
Academic Qualifications	High school and below	50.0	50.0	15.2
	College or undergraduate	31.5	68.5	68.4
	Postgraduate and above	15.4	84.6	16.5
Health Condition	Health	36.4	63.6	69.6
	Sub-health	18.2	81.8	27.8
	Disease	50.0	50.0	2.5
Net Annual Personal Income	$50,000 and below	45.9	54.1	46.8
	$50,000–$80,000	30.0	70.0	12.7
	$80,000–$120,000	14.3	85.7	17.7
	$120,000 and above	16.7	83.3	22.8
Is there an old person in the family?	YES	31.9	68.1	91.1
	NO	28.6	71.4	8.9

Annotations: $^{*}p<0.05$; $^{**}p<0.01$; $^{***}p<0.001$

higher academic qualifications. However, the survey revealed a lack of consumer awareness regarding *Cordyceps sinensis*, with only a small number of individuals capable of distinguishing its authenticity and efficacy. A mere quarter of those surveyed had prior purchasing experience, and for these people the consumption recommendation issued by the CFDA had a limited impact on their consumption of Cordyceps as a medicine but affected its consumption as food.

Regarding awareness of the potential arsenic risks, 73.3% of respondents were unaware of the consumption guidelines issued by the CFDA. This suggests that only 12.2% of consumers were aware, suggesting the guidelines were ineffective. Initial cross-analysis of factors influencing consumer awareness of arsenic risks indicates that most demographic characteristics did not have a significant bearing on arsenic risk awareness. However, gender plays a significant role, with men exhibiting greater familiarity with arsenic contamination risks compared to women. Age, education, health status, personal annual net income, and the presence of elderly individuals in the household did not demonstrate statistically significant differences in awareness; however, consumers who had previously purchased dried *Cordyceps sinensis* or processed products exhibited higher awareness of arsenic contamination, particularly among those who had consumed Cordyceps over an extended period.

Given the limited understanding of the arsenic contamination risk among consumers, we conducted an experiment to assess their purchasing intentions. The results indicated that when consumers became aware of the arsenic contamination issue and the associated consumption guidelines, their willingness to purchase *Cordyceps sinensis* significantly decreased, with an approximate 85% decrease in their willingness to recommend the product. Further cross-analysis of consumer purchasing behavior revealed that age was statistically relevant to the willingness to purchase, with older consumers demonstrating a greater decline in their interest to purchase after becoming aware of the arsenic issue. Gender, education, health status, personal annual net income, and the presence of elderly individuals in the household did not exhibit statistically significant differences. Among consumers who had prior purchasing experience, nearly 68.8% were still willing to continue buying it. Approximately 22.2% were not willing to explore long-term Cordyceps consumption but were still open to consuming it. These findings suggest a statistical correlation between long-term Cordyceps consumption and willingness to continue purchasing the product. Despite the concerns raised by the arsenic risks, there remained a segment of consumers who perceived *Cordyceps sinensis* as a valuable product and were willing to explore alternative ways of incorporating it into their lifestyle.

It is crucial to educate consumers about the risks associated with long-term consumption and to establish clear guidelines to ensure the authenticity and quality of Cordyceps products on the market (Liu, 2013; Pang, 2014; Wang et al., 2016). By doing so, consumer confidence can be restored, leading to a healthier and more sustainable Cordyceps industry.

To summarize, the debate over the arsenic content of *Cordyceps sinensis* has shown a strong need for the continuation of scientific research into the biochemical qualities of Cordyceps and for regulating the consumption of associated products. Regulating authorities and industry associations could do more to guide their members—the suppliers of Cordyceps—who deliver processed products for consumers, taking into account that some of these consumers will have chronic conditions. After the debate, it was established that Cordyceps companies should monitor arsenic content and that consumers should limit their consumption to 4 grams per day or to 5 months for oral drug administration (Liu et al., 2018).

It is also curious that after watching the HBO show 'The Last of Us' (2013), named after the video game, some Western consumers have been afraid of consuming Cordyceps. Several mycologists have spoken in the media to destroy anti-scientific beliefs and to cool down consumers' anxiety (McNeal, 2023).

References

Au, W.Y. (2011). A biography of arsenic and medicine in Hong Kong and China. *Hong Kong Medical Journal*, *17*(6), 507–13.

Cao, X., Wang, J., & Li, J. (2015). Analysis of arsenic compounds in *Cordyceps sinensis* of Xizang by HPLC-HG-AFS method. *Chinese Patent Medicine*, *37*(9), 1985–1989.

Chen, P.X., Wang, S., Nie, S., & Marcone, M. (2013). Properties of *Cordyceps sinensis*: A review. *Journal of Functional Foods*, *5*(2), 550–569.

Cui, J. (2019). *Research on Earnings Management Behavior of 'ST-Cap Off' in Qinghai Spring (600381)*. Donghua University.

Department, D.M.S. (2019). *Cordyceps sinensis*, capital chase under a medical farce. *Golden Era*.

Giri, S., & Singh, A.K. (2016). Spatial distribution of metal(loid)s in groundwater of a mining dominated area: Recognising metal(loid) sources and assessing carcinogenic and non-carcinogenic human health risk. *International Journal of Environmental Analytical Chemistry*, *96*(14), 1313–1330.

Jiang, Z., Xu, N., & Liu, B. (2018). Metal concentrations and risk assessment in water, sediment and economic fish species with various habitat preferences and trophic guilds from Lake Caizi, Southeast China. *Ecotoxicology and Environmental Safety*, 157, 1–8.

Khan, S., Cao, Q., & Zheng, Y. M. (2008). Health risks of heavy metals in contaminated soils and food crops irrigated with wastewater in Beijing, China. *Environmental Pollution*, *152*(3), 686–692.

Liu, J., Li, Y., & Zan, K. (2016). Comparison of 5 kinds of heavy metals and harmful elements in artificial and wild *Cordyceps sinensis*. *Chinese Pharmaceutical Affairs*, *30*(9), 912–918.

Liu, L., Zhang, Y., Yun, Z., He, B., Zhang, Q., Hu, L., & Jiang, G. (2018). Speciation and bioaccessibility of arsenic in traditional Chinese medicines and assessment of its potential health risk. *Science of Total Environment*, 619–620, 1088–1097.

Liu, Y. (2013). Key points of identification of *Cordyceps sinensis* and its counterfeit and miscible characters. *Inner Mongolia Traditional Chinese Medicine*, *32*(30), 86–87.

Maghakyan, N., Tepanosyan, A., & Belyaeva, O. (2017). Assessment of pollution levels and human health risk of heavy metals in dust deposited on Yerevan's tree leaves (Armenia). *Acta Geochimica*, *36*(1), 16–26.

McNeal, B. (2023, February 5). We asked a mycologist about The Last of Us and it got weird. *Esquire*. https://www.esquire.com/entertainment/tv/a42760795/last-of-us-fungus-cordyceps-mycologist/

Pang, L. (2014). *Research on marketing strategy of 'Cordyceps' high-end cigarette brand based on consumer psychology*. Inner Mongolia University.

Sanders, A.P., Claus Henn, B., & Wright, R.O. (2015). Perinatal and childhood exposure to cadmium, manganese, and metal mixtures and effects on cognition and behavior: A review of recent literature. *Current Environmental Health Reports*, *2*(3), 284–294.

Shao, J., Hu, Y., & Liu, S. (2020). Morphological analysis and evaluation of arsenic in Cordyceps. *Chinese Journal of Analytical Sciences*, *36*(2), 229–234.

Shen, J. (2018). *Cordyceps sinensis* removed as health care product. *Chinese Quality Travels Ten Thousand Miles*, 8, 38–39.

Wang, H., Shan, Y., & Sun, Z. (2016). Application and market status analysis of *Cordyceps sinensis*. *Modern Chinese Materia Medica Research and Practice*, *30*(6), 83–86.

Xiao, M., He, M., & Tang, C. (2012). Research hotspot and frontier analysis of *Cordyceps sinensis* based on CiteSpace Knowledge graph. *Chinese Journal of Microbiotics*, *42*(12): 107–108.

Xiao, Y., Li, C., & Yang, H. (2021). Research progress of arsenic in *Cordyceps sinensis* and its safety evaluation. *Chinese Herbal Medicine*, *52*(15): 4731–4740.

Yuan, Y. (2016, February 26). Can you still eat Cordyceps? *China Science Daily*. http://news.sciencenet.cn/htmlnews/2016/2/339054.shtm

9 Rural Land Reforms and Management of Cordyceps in Qinghai

Lessons for Sustainability

Ksenia Gerasimova, Jiang Zhao, and Jiping Sheng

Introduction

A brief review of countries where there are wild *Cordyceps sinensis* pastures can show that different types of land ownership and management of the highland grasslands where Cordyceps are harvested does not fully prevent overharvesting, which is a threat to its conservation. Understanding the roots of the institutional complexity of the current models of land management can serve as a foundation for organizing more sustainable harvesting and securing this species' survival under even more challenging climatic conditions in the near future. Our discussion starts with a reference to the Common Pool Resources dilemma debate and then moves to the detailed consideration of the evolution of institutions managing Cordyceps harvesting, which is illustrated with relevant case studies collected from Qinghai Province. We explore how entitlement to Cordyceps collection and harvesting practices has changed under the Chinese rural land management reforms and the effect on Cordyceps' ecology as a result. These findings are crucial for ensuring sustainable management of Cordyceps habitats in the future.

Theoretical Background

While the type of property rights regime is not the single factor determining the ecological threshold of natural resources of the Qinghai-Tibetan grasslands—other socio-economic as well as physical and climatic dimensions should be added for a comprehensive assessment—it is still one of the most important drivers defining the level of resource consumption, as it sets the rules of access to the resource and can provide dis-/incentives that influence the type of collecting behavior of key actors.

Originally, discussion of the Common Pool Resources (CPR) dilemma came from a discussion on rural land management. Garrett Hardin in his 1968 influential essay described the tragedy of 'Freedom in a Commons,' using the example of a pasture which was overused by herdsmen in pursuit of their individual interest and how the long-term negative effects of this overburdening depleted the land (Hardin, 1968). And the essence of the sustainability dilemma itself is how to provide for the needs of the present population without compromising resources for the generations to come, isn't it?

DOI: 10.4324/9781003429753-13

This influential argument has been extended to include the debate on the role of the property rights regimes. In the first update of his theory, Hardin identified the core of the tragedy of the commons as 'unmanaged' access to resources (Hardin, 1994). Theesfeld (2019) suggested four possible models of natural resource management to resolve this tragedy, using two key factors: level of centralization vs local engagement and individual vs collective interests (Table 9.1).

As one can see, the two extremes (A and D) include management by a single central authority and a fully decentralized system based on individual privatization, and the two models in between require high levels of engagement within local systems (B and C).

Hardin's original response to the tragedy of commons was to privatize the land or the right to access it (Hardin, 1968). Under Coase's property rights theory, privatization of public resources should reduce externalities and provide people with incentives to internalize negative externalities and produce efficient economic operation output (Coase, 1960).

However, in the second revision, Hardin recognized that if resources are scarce, the trap of the tragedy of commons will remain: 'Under conditions of scarcity, ego-centered impulses naturally impose costs on the group, and hence on all its members' (Hardin, 1998).

Ostrom rejected the individual privatization model, since 'individuals who have high discount rates and little mutual trust act independently, without the capacity to communicate' and 'are not likely to choose jointly beneficial strategies unless such strategies happen to be their dominant strategies' (Ostrom, 1990, p. 183). And the model based on full control by the central government is most likely 'to destroy institutional capital that has been accumulated during years of experience in particular locations' (ibid., p. 184).

Independent self-organized grassroots institutional management is well-illustrated by empirical cases in the literature (ibid.). The management system organized by a public manager, who is 'a civil servant directed by the community,' is different from the 'extreme' central state management model. The relative success of the intermediate models over extreme ones seems to rely on a higher level of engagement of local communities, which, in its turn, requires additional policy measures, such as community consultations and education (Theesfeld, 2019, pp. 345–346).

The case study of the institutional management of Chinese grasslands producing Cordyceps, which we discuss below, provides more support for this argument

Table 9.1 Models of resource management to overcome the CPR dilemma

Centralized		*Localized*	
A. Fully controlled by central government	B. Run by the public manager	C. Independent grassroots collectives	D. Fully individualized decision-making
Collective rights		**Individual rights**	

by showing the inefficiency of the fully centralized management model, but it also reduces optimism concerning the intermediate models' efficiency. The case findings show that institutional lock-in effects are not easy to overcome and require much effort from policy-makers as well as high environmental awareness from the local community, plus time for institutional change to settle in. It also contributes to the discussion of the endogeneity and exogeneity of institutions managing natural resources.

Methods and Materials

The chapter employs an ethnographic methodology based on a qualitative research paradigm. Our ethnographic case study approach was to merge primary data collection and study of secondary ethnographic materials from earlier ethnographic expeditions.

Cordyceps are endemic to the Tibetan Plateau, thus the experience of collecting Cordyceps is rather unique and confined to small segments of the rural population of Bhutan, China, India, and Nepal, and for the rest of the world, these mysterious fungi are still an exotic oriental product. Even though there are general trends such as rapid rises in the price of Cordyceps and climatic pressure affecting the amount and quality of Cordyceps available, different national sites might produce different localized experiences. Thus, we have focused on only one ethnographic site, the Cordyceps-producing areas of Qinghai Province.

We conducted our own fieldwork study in Xining city in May 9–10, 2018 and complemented the material obtained from observations and interviews conducted at the famous local Cordyceps market with interviews from earlier studies in Qinghai conducted by Chinese ethnographers such as Fan Chang-feng (2016) and Cai Bei (2014), which is the most common ethnographic method combination (Seligmann & Estes, 2020). All three studies have revealed similar patterns in the local Cordyceps-processing industry, and we were able to identify how changes in the patterns of Cordyceps collection have followed the course of Chinese rural land reforms.

The chosen methodology is qualitative and focuses on understanding unique individual experiences of the Qinghai rural population engaged in wild Cordyceps collection. To maintain their voices, we present extracts from interviews which are the most illustrative of specific patterns of pasture management and collection and trading of unprocessed Cordyceps. Thus, individual experiences are collected in a geographical case study, which is then used to discuss universal problems of fairer resource distribution and sustainable management of natural resources.

Rural Land Reforms and Harvesting Cordyceps in Qinghai Province

In 1978, China introduced a set of reforms aimed at transforming its rural land tenure system, in terms of a transition from collective land ownership and management to the new household contract responsibility system (HCRS), in order to gradually move into free trade of rural land (Xu, 2016; Deininger & Binswanger, 1999). The new aspiration for the establishment of a market economy was complemented with securing the 'social justice principle' and public ownership of land (Ho, 2013). Under the HCRS, ownership rights were given to the rural collective and the contractual and use rights to farmers. Before 2002 the contractual rights had not been secured; communal cooperation in pastures had been disrupted and environmental degradation was increasing due to overuse and natural hazards (Li et al., 2018).

Since the Chinese government withdrew a complete privatization model, a new reform has been developed under the name of 'separating three property rights' (STPR). It divides rural land rights into three components: non-tradable ownership, non-tradable contractual rights, and tradable land use rights (ibid.).

As for Cordyceps, which are the produce of Tibetan highland pastures, the original land reform focused on grazing rights and not on Cordyceps harvesting but still brought changes to how and by whom wild Cordyceps were collected. The timeline in Figure 9.1 shows the periodization of the Chinese rural reforms in regard to ownership and management of pasture highlands, which identifies three distinct periods.

Despite the first stirrings of reform back in 1978, actual implementation started much later, as until 1996 it was still the period of public ownership of the highland pastures. The first target of granting households the right to graze livestock on land was completed in the first five years after the beginnings of reform, which was then followed by allocating individual households use rights to winter-spring pasture in the period between 1997 and 2002.

However, Cordyceps collection rights were still controlled by the rural collective, since the harvesting time, depending on the altitude, was in the summer-autumn. Slowly, from 1997 onwards, land use rights for summer-autumn pastures began to be institutionalized, which lasted until 2003. After 2003, the new year-long pasture contracting system was implemented (Awangjiancuo, 2004). The process coincided with the rising demand in Cordyceps and subsequent price increase, leading to a high influx of people entering the area to gain profit from collecting Cordyceps, but the target was to adopt a reasonable management model that protected natural resources and benefited local poor people; to get rid of poverty became a new agenda.

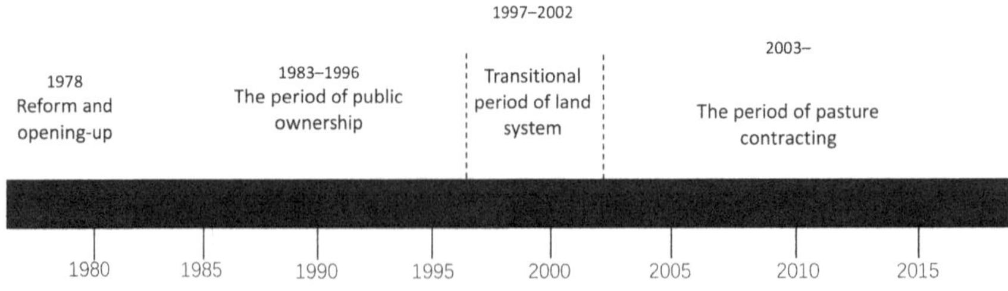

Figure 9.1 Land management reform in Qinghai Province

The Period of Public Rural Land Ownership (1978–1996)

At this stage, the household contract responsibility system was gradually introduced to the eastern part of China. However, in the Qinghai-Tibet region, public land ownership still remained, meaning access to land was owned by the village community rather than individual households. In 1983, the local government of Qinghai proposed 'The Tentative Measures for the implementation of the responsibility system of the livestock contract to the household' (<关于实行牧业包干到户责任制若干问题的试行办法>). However, during the years from 1983 to 1992, the land system reform in the region was not complete, and only the goal of granting livestock grazing rights to households was achieved; the land was still under public ownership (Fan, 2016).

In 1992, the government of Qinghai Province proposed 'Opinions on further stabilizing and perfecting pasture household contract system' (关于进一步稳定和完善草场承包制的意见). In 1996, winter-spring pasture rights had basically been achieved, but autumn-summer pasture rights had not yet been institutionalized (Awangjiancuo, 2004). As Cordyceps mainly grows in high-altitude summer-autumn pastures and winter-spring pastures are mostly low altitude, the reform had actually very little to do with Cordyceps. Besides, the price of Cordyceps was not high at that time, and the locals chose to dig only small quantities of Cordyceps for extra household income while keeping livestock. There was no limit for harvesting Cordyceps, so everyone was able to step into pastures to collect them. This individual economic behavior is referred to as 'separate digging,' meaning that the collection was unorganized (Fan, 2016).

The method of trading Cordyceps was as simple as trading any other items in Tibet. It was mainly sold by Tibetan herdsmen to the Hui people for extra income. Without land transfer (tradable land use rights), it was impossible to develop Cordyceps collection and trading into the market economy model. As there was no private ownership, Cordyceps were regarded as a public natural resource, and the species suffered from excessive harvesting, since there was no one to control the collection volumes (ibid.). Below is a records of that period according to a Cordyceps collector and a trader.

Case 9.1. Cordyceps Trader in Linxia, 'ZSF,' Male, 50 Years Old

I have been doing Cordyceps business in Qinghai Province for 22 years. From 1982 to 1992, it was my start-up period of working as a medical drug dealer. Every year, I came to the pastoral area to purchase Cordyceps from the locals from April to June. It was about 0.7 or 0.8 yuan for each piece with a few cents of profit. When you came across good luck, you could earn 0.1 yuan. It took five or six days for each visit to the pastoral area. Two or three persons could get 5–10 million pieces at one time and make a modest profit. When the price of Cordyceps was higher in 1996, I opened a specialty store in the Xinghai

County and lived in the Huangheyuan Hotel. At that time, there was no public Cordyceps market in Xinghai.

In the 1990s, the amount of people coming to Tibetan areas began to grow. Sometimes, there were hundreds of people in a grassy mountain who were digging up Cordyceps. At that time, the resource was still rich. If one person dug 100 pieces one day with each worth 1 yuan, it would be much better than going out for other work. At that time, people basically had no awareness of ecological protection, and they did not backfill them. The Tibetans were better. The condition was not highly regulated by the government, and the ecological damage was rather serious.

Source: Fan, C. (2016). Economic form and cultural change of *Cordyceps sinensis* in Qinghai-Tibet area. *Folklore Studies*, 1, 118–128.

Case 9.2: Local Herdsman in Golog, 'HD,' Male, 41 Years Old

When I was a child, I dug Cordyceps everywhere. There was no concept of private property at that time. Unlike people saying now 'this is my mountain; you cannot dig here.' We would go wherever we wanted to find Cordyceps. The price of Cordyceps was quite low and the highest was about 1.7–1.8 yuan for one piece. It was regarded as a sideline. Because we cannot live totally on that. And there were no outsiders coming here to dig Cordyceps.

Source: Cai, B. (2014). Trading patterns of *Cordyceps sinensis* in Tibetan Regions—Take 'floating' *Cordyceps sinensis* of Golog Tibetan Prefecture of Qinghai as example. *Qinghai Journal of Ethnology*, 1, 132–135.

Transformation of the Rural Land System (1997–2002)

In 1997 Qinghai local authorities, for example in the area near the Animaqingshan Mt, ranked the 23rd highest in China, started granting summer-autumn pasture rights to local households. By 2002, households had full pasture rights (Awangjiancuo, 2004).

During the transitional period, the price for Cordyceps went up and more people poured into the mountainous grasslands for digging; access was still managed by the community. As shown in Figure 9.2, people who intended to engage in the extraction of Cordyceps had to pay for the certificate. The income from contracting Cordyceps collection was distributed to governments at all levels. The county-level Cordyceps Office, the Pasture Station, and the Animal Husbandry Bureau were responsible for printing certificates and stamps. The community group was responsible for collecting so-called digging fees and issuing digging certificates. The

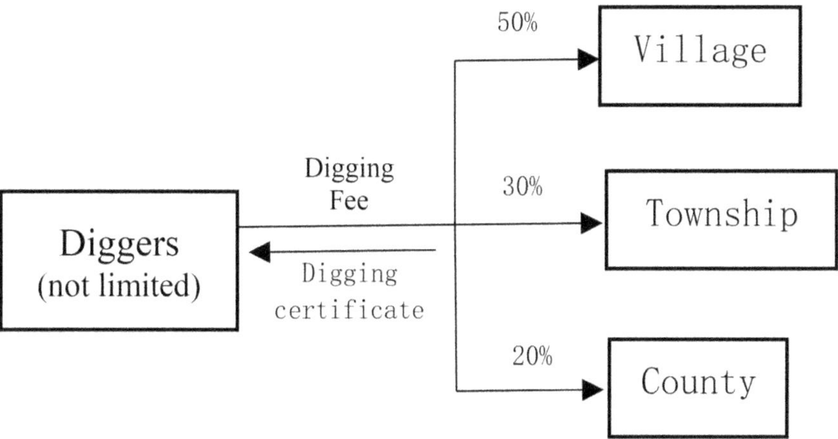

Figure 9.2 The relationship between subjects during the transition period

villagers were responsible for patrolling mountains and supervising. Individual families had basically nothing to do with the process (Fan, 2016).

At this stage, under the inertia of public ownership, the herdsmen as landowners still did not know how to deal with the ever-changing Cordyceps phenomenon, and the local government was also confused in the face of increased harvesting. In terms of the distribution of benefits, the profits from contracting Cordyceps collection were shared between the county and the villages, while the herdsmen who owned contracted land benefited little. From 1997 to 2003, the proportion of benefits indicated that the various departments had profited a lot from Cordyceps resources (Fan, 2016). In other words, it made sense to accommodate as many collectors as possible to obtain more collection fees, which went into the government's budget, but competition from outsiders and the pressure on the ecological system increased.

Case 9.3: Local Official, 'ZBJ,' Male, 38 Years Old

The opening up period started from 1996 or 1997. Outsiders began to enter into this area to dig Cordyceps. Before that, the price was very low; only the locals dug on a small scale. For the locals, Cordyceps was a kind of Chinese medicine. People bought it to treat diseases. After 1997, the price stated to rise to 1 yuan for each piece, which was very attractive because at that time laborers working in urban areas could only make 10–15 yuan/day. Cordyceps could bring 20–40 yuan/day conservatively. Sometimes, one could even reach 100–200 yuan/day at the highest. As a result, many people came here to dig Cordyceps.

Xueshan town, Dongqinggou town and Dawu Town in Golog County have had the richest resources of Cordyceps and large areas of pasture. When the 'Cordyceps army' came, the mountain gully was full with tents, like celebrating an important festival. Although the pasture was already contracted to households, the right of management was still owned by the community group in a local town. And the income from Cordyceps was distributed among the county, town and village. And the ratio was 20%, 30% and 50%, respectively. Individual or contracted households had no benefit from that. Although there were institutions like the Cordyceps office, the pasture management station and the Animal Husbandry Bureau in local government at that time, the community group had the central power over who had contractual rights to dig Cordyceps. In other words, the decision was made by a community group on how many people were allowed access to the contracted household pasture. And it was also the community group that charged the digging fee. At that time, the fee was 400–500 yuan/person. Upon receiving the digging fee, the community group would provide a digging certificate for diggers with the seal of the community group. It was still profitable to dig Cordyceps after paying the digging fee because a person could dig 1 kg of Cordyceps on average, valuing 2,000 yuan. With the 500 yuan digging fee, there was still 1,500 yuan profit. The main problem was that large amounts of people caused serious destruction to the pasture land. People were not conscious of protecting the pasture. They were not aware of filling the pit after digging. There were patrols (a horse riding team) in each village to check digging certificates. If a person was found that had no digging certificate, he would be withdrawn. And the Cordyceps he had dug would also be confiscated. People were honest at that time; seldom people faked the digging certificate. This regulation did not bring benefit to local farmers (contracted householders). All the income from digging fees went into the public purse. Part of it was used to repair the school in a community group, some was used to take care of poverty reduction. However, some money went to individuals—into the leaders' pockets, like the secretary of the community group, the village leader and the accountant of the village.

Source: Cai (2014).

Case 9.4: Digger who Pays Pasture Fee in Xunhua, 'ZD,' Male, 37 Years Old

At the beginning, I mainly went to Dari, the Gande area, and did not go to Golog (the Maqian County). At that time, I had to be led by the boss because I was not allowed to step onto the estate. At that time, there were a lot of Cordyceps. We usually carried our luggage and took the shuttle to Xining.

Then I had to find a truck, which was a little difficult to get. And it was more than ten years ago when the fee was 60 yuan. The whole journey took usually 3 days. I would get something to eat when I got to a nearby town. Then I would go to look for the boss or the digging pasture leader to find Cordyceps. The community group was in charge, which means you had to rely on friends and relatives to get in touch to see if you could dig Cordyceps on the pasture. The admission fee to the pasture was also expensive, and the highest was 7,000 yuan and even 8,000 yuan. At that time, half a kilogram of Cordyceps was as high as 2,000 or 3,000 yuan. The first time I went to Golog, I went to the east of the Mazhi County. A place called Zhelang was also a place with Cordyceps. The digging fee was 8,000 yuan, which was paid separately in 2 installments. I was to pay 4,000 yuan after paying the first 4,000 dollars. I lived in a local household and ate with them.

At that time, it was very difficult. I had to cross the river and go at night because there were interceptions on the road. Some people rode motorcycles, but walked when they came to the barrier. Most people just walked. Without a big way to walk, everyone was sneaking. I have not ever been caught. We dug well, that is, a pound and more than three or two, one person a month. Except for the pasture fee I had paid, we were able to earn more than 20,000 yuan. Such good conditions lasted for two or three years. It was not that lucky after that.

Source: Cai (2014).

New Land Contracting System (2003–)

The influx of new labor had a huge impact on local social order and public security. In 2003, the government proposed the policy of ‘prohibiting external cordyceps collection and limiting internal collection,’ which meant outsiders were not allowed to dig Cordyceps on the Qinghai pastures (Cai, 2014).

Local people could pay digging fees for digging certificates with their registered permanent residence permits. With digging certificates, the locals were able to collect Cordyceps anywhere under the jurisdiction of the Pastoral Committee. There were check-points or ‘barriers’ on the way up to the pastures. Armed police forces were responsible for patrolling the mountains.

At that time, there were lots of cases of people borrowing or forging household registers. And the illegal digger would have to take the local people’s car in order not to be arrested and sent back behind the barrier. Everyone knew that they could get away with it by paying more. Local people took advantage of the situation to earn more income by helping outsiders to borrow household registrations and pass the checkpoint—there were still lots of outsiders attracted by huge profits. Diggers, businessmen and local staff were involved at various stages of contracting Cordyceps collection, as summarized in Figure 9.3.

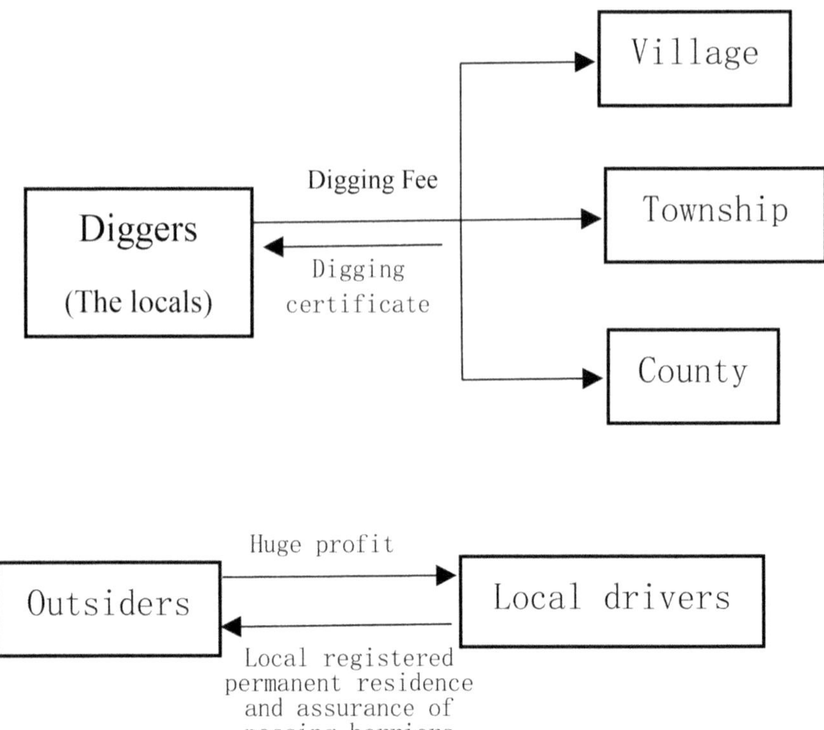

Figure 9.3 The relationship between subjects during the pasture contracting period

Case 9.5: Local Government Staff Member, 'ZBJ,' Male, 38 Years Old

At that time, town and county worked together to prohibit outsiders digging Cordyceps in this area. Only the persons who were local (with registered permanent residence permits) had access to dig Cordyceps in Golog. There was a Cordyceps office in the pasture supervision station in the county. People who wanted to dig Cordyceps in this county had to go to this office to apply for the digging certificate. There was a specified place to pay the digging certificate fee (including tax). Usually, a digging certificate would be 2,000–3,000 yuan, sometimes even 5,000–6,000 yuan. On the digging certificate, you could see the information about the registered permanent residence permit and also the photo of the digger. There were a few fake digging certificates at the beginning. Some outsiders who borrowed the permits could also enter into this area. And drivers or Cordyceps businessmen would help them with the papers. Gradually, people began to fake the permits. Each fake permit would be charged 100 yuan. If a household covered outsiders, they would be punished. Even the hotels didn't accept guests outside the state. Armed policemen were ordered to patrol in this area. There were often conflicts between the army and the outsiders. Some were even shot by the soldier. So the enforcement was very strong. When outsiders who had dug Cordyceps were found, they would be taken outside this area.

Source: Cai (2014).

Case 9.6: Local Tibetan Driver, 'ZX,' Male, 43 Years Old

There was a time when people who had registered residence permits in Golog were able to dig Cordyceps. So, we usually spent 100 or 200 yuan to borrow more than 100 permits. Then we gave them to the people who were to dig Cordyceps. A car full of diggers had to pass barriers on the way to pasture. You could pass it if you had someone you knew, or you would be punished.

The pasture which you had owned for two or three years would be taken away by others. This was the biggest fine. The small fine was a penalty. I used to take approximately 120 or 130 persons in a Dongfeng big truck. You know, we were usually familiar with people at the barrier, so generally, we talked to the leader in advance, as well as other persons. We often called when we were about to reach there and described what our car was like. A car full of persons was covered well. When it was asked what was in the car, we lied and said it was cement or bricks. And then a person just pretended to have a look at it and went back. Generally, this was the case. Sometimes we bought something to eat for the people at the barrier, such as meat and drink; we rarely gave them cash. At that time, the price was high to take people from Xining to Golog as there were many checkpoints on the way. Sometimes we passed Huarixia and Gande, also Xiadawu and Lajia. The charge ranged from 800 to 500. It had been three years when I was able to earn about 200,000 yuan one year by doing this. At that time, there was no land for sale (the pasture was sub-contracted one year), and there was no hired digging at that time. It has been three years since they began to hire diggers and now, in the current year [2012], it is also the same for selling land. For the last two years I haven't worked as a driver any more. You know, you won't earn much without the checkpoint set at the barrier.

Source: Cai (2014).

Case 9.7: Truck Driver in Malang, 43 Years Old

It was impossible to prohibit outsiders from digging Cordyceps completely. First, digging required labor, and secondly, there would be more pasture fee income if there were more diggers. There were two reasons why outsiders couldn't come in. One was they didn't have local registered permanent residence and the other was the checkpoints restricted relevant vehicles. I was a local truck driver. As I often pulled people or goods, I gradually got to know people on the barrier. When diggers chartered the car, I borrowed plenty of household registration permits from Maduo locals and lent them to diggers, for which I charged one or two hundred yuan. My Dongfeng truck could accommodate one hundred and twenty persons at most. According to the distance, the fare charged was between 300–800 yuan, which was much more expensive than the passenger car. This is mainly because the barriers must be solved. Before I set off, I would call the person at the barrier and tell the person in charge what type, color and license plate number; they told me to just hide people and answer that the truck is pulling cement or bricks and so on. Besides, I generally left them two sheep or something to eat and drink. In recent years, the situation has changed. Without the restrictions of registered residence and checkpoints, the fee is not high anymore. I do not run the car now.

Source: Fan (2016).

When administrative measures could not control the influx of the growing digging army, the government realized that they had put under threat herdsmen's rights to use and protect pasture, which was determined by the new pasture contracting system. The government at all levels withdrew from the practice of competing for profit with the contractual rights owners.

As Figure 9.4 shows, the three-level co-management became a two-dimensional model of village-household co-management, which meant the households could get 80% of the pasture fee, and the remaining 20% revenue was owned by the village.

At that time, the government's engagement concentrated on provision of public services such as public security, ecological security, and weather forecasting. Herdsmen had the right to inherit their pasture and protect it from intruders. For outsiders, the Cordyceps resources were no longer 'wild and [with] no owners,' nor were they collectively owned by the community but the land owners (Fan, 2016). This greatly reduced entry for external diggers. At the same time, families with access to Cordyceps pastures could benefit from it. However, most herdsmen were still rather poor.

After 2008, the pasture contracting system had been fully completed and implemented, which meant that the planned land ownership transfer became a reality, and the market system gradually matured. Social management shifted from collective

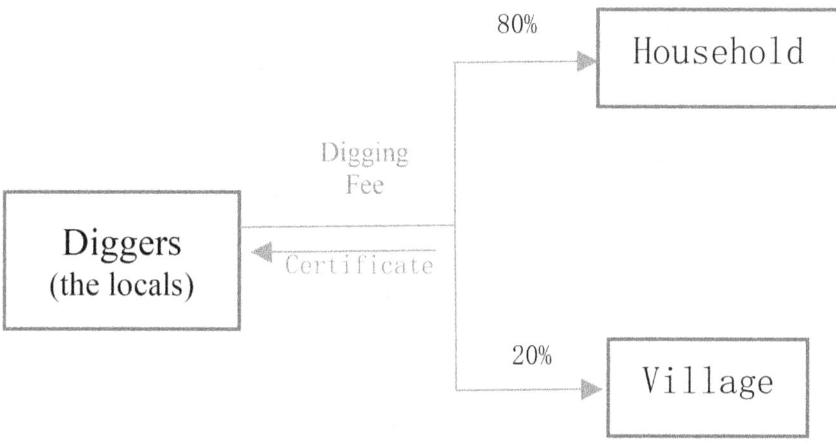

Figure 9.4 The relationship between subjects during the period of village-household co-management

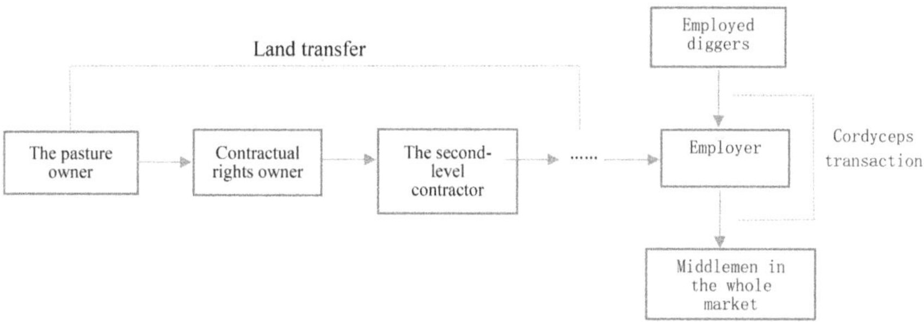

Figure 9.5 The relationship between subjects in the secondary market during the pasture contracting period

Source: Ho (2013).

management and barrier interception to more rational administrative guidance, and pasture control was finally given to herdsmen.

At this point, a new way of trading Cordyceps was born—that is, most pastures were contracted out, and the contractors transferred pasture rights once again or hired migrant workers to dig Cordyceps to make profits. Two types of contracts were promoted: one was called a 'pasture contract' signed by 'Baoshan boss' (the village head) and the owner of pasture, and the other was the contract for the excavation of Cordyceps, signed by the boss and migrant workers (Fan, 2016).

This new trading model also benefited the pasture owners, as they became the first-level contractual rights owner. The person who does the contracting on behalf of the owner became 'the second-level boss.' After the land was transferred, it produced a 'third-level boss,' 'Fourth-level boss' and so on. Finally, 'the last-level boss' who contracted the pasture would hire workers to dig Cordyceps. The employers didn't have to pay for their board (Cai, 2014).

Generally, the diggers employed would sell Cordyceps to 'the boss' at a specified low price. Then the boss would negotiate sales with buyers on the Cordyceps market for profit. Finally, the owners of pastures started to act as the employers of diggers. However, as the harvest of Cordyceps in the pasture is often affected by various natural factors such as precipitation, it also meant taking higher risks in contracting pasture. In contrast, the hired workers did not risk losing money as they negotiated the payment in advance. Thus, in this contractual rights management system, the hiring boss, driven by profits, would often employ an excessive number of diggers, which posed a serious threat to the habitat.

Case 9.8: Businessman A in Jiuying Market, Male, 53 Years Old

Businessman A runs a store in the Jiuying Cordyceps market in Xining City, Qinghai Province. He is Tibetan, from Haidong, Qinghai Province, and was born in 1965. He had been engaged in the Cordyceps industry for 26–27 years at the time of interview. The company mainly sells Cordyceps, excluding processed products such as lozenges. In addition to Cordyceps, the company also does business in other agricultural products, such as black mites.

My company makes purchases of Cordyceps through two major means: the first one is by contracting pasture, and the other one is to buy Cordyceps from the local farmers and herdsmen instead of the wholesale market. Since 2005, the company has been contracting pasture with Cordyceps. In 2017, the company signed for two grass mountains in Golog, with one 300,000 yuan/year and the other one 500,000 yuan/year. You are generally required to pay 20–30% of the rent as a deposit; it varies with different herdsmen. And the remaining rent is paid after the gathering period. In 2017, the company employed 50–60 people to dig Cordyceps. The workers are paid on a piecework basis—that is, 7 yuan for each piece, and the wages are also settled at the end of the gathering season. The most important criterion for selecting diggers is whether they can gain the trust of the boss—that is, they can assure the boss they won't hide Cordyceps. In 2017, nearly 60,000 pieces of Cordyceps (20–30 kg, 3,000 pieces/kg) were harvested from the mountains, and the output was quite satisfactory. In 2018, the company expanded further to contract four mountains. And the rent for the pasture rose a lot as the output was relatively high last year. It can be seen that the price has risen to 1.3 million yuan with an increase of 30%. Therefore, the company suffers from high cost and risk. In 2017, excluding the cost of marketing, circulation and management, the price of Cordyceps reached 20 yuan for each piece only, considering the investment for labor and land.

I would go to the base every year to purchase Cordyceps for two reasons. First, the price of Cordyceps is lower than that of Cordyceps on the market. Second, the quality of Cordyceps dug from the bases is better. I am very familiar with farmers and herdsmen, and build trust with them. I won't purchase at a higher price than the market one. Purchasing prices and quantity are decided according to the market situation, sometimes 10,000 pieces one day, or 20,000.

When it comes to the impact of the Internet, the popularity of information technology has accelerated the speed of information dissemination, whose main medium is WeChat. The use of WeChat means the fake Cordyceps have nowhere to go. For example, some time ago, when white fake Cordyceps appeared on the market, news spread so quickly in the WeChat circle of friends that dealers as well as buyers all got to know immediately. Therefore, the fake Cordyceps could be expelled from the market soon by the supervision department of the Jiuying wholesale market. Second, the use of WeChat has increased the number of new customers. With the spread of WeChat, it is easier for major customers to discover us and establish direct contact. Before that, we had difficulty meeting some big customers as they may have been targeted by other businessmen. For example, this year [2018], Qiaorui Agriculture in Shanghai purchased 4,000 pieces of Cordyceps from us for the first time. They believed in us because we had our own bases for a long time, and Cordyceps were all purchased from farmers and herdsmen. For large customers, it is beneficial to contact us directly. Not only is information about the quality of the products uncertain but also the price is relatively high, as the two traffickers need to pocket the difference from the transaction.

Source: Our research group interviewed traders in Qinfen Lane, Xining, in May 2018.

Case 9.9: Mobile Vendor B in Jiuying Market, Male

B is a mobile vendor in the Jiuying Cordyceps market. He is of Hui nationality, born in Huangzhong County, Qinghai Province in 1968. He has been trading Cordyceps for more than 20 years, and his supply source mainly comes from the pasture he signed for. When the supply is not enough, Cordyceps is purchased from the wholesale market. B's business is mainly run by his family. His son is responsible for the base and online shop, while B and his three brothers go to the market in Xining (he pays his brother a salary in lieu of a share of the business).

If sales are good, I will give each brother 2–3 million yuan. If the sales situation is just normal, the salary for each is more than 10,000 yuan, which is quite ideal for the locals for 40 working days. Three persons have to cooperate with each other. As fresh Cordyceps come into the market every day during the digging period, one of them must reach the base every day to transport fresh Cordyceps to the wholesale market in Xining. Generally, it takes up to 12 hours to get from Xining to Yushu by bus. In order to stay on top, we often sell a batch of Cordyceps in the morning then take a bus to the Yushu base at noon, and arrive at night. After getting Cordyceps, we take the car overnight to get back to the Jiuying Cordyceps market before dawn.

In 2017, I rented pasture in the Yushu area from the local herdsmen. In my view, the quality of Cordyceps harvested from Yushu is better than Golog because of the higher altitude; the proportion of Cordyceps with yellow eyes is higher. The rent of the pasture was 480,000 yuan in 2017, and the deposit was fixed at 200,000 yuan.

After the excavation period, the full payment is made to the herdsmen directly. I have been contracting pasture in Yushu since 2011; the rent has risen rapidly from more than 100,000 yuan in 2011 to more than 200,000 yuan per year [2018]. The reason for the increase is that the price of Cordyceps has been increasing year by year. In 2008, the price of Cordyceps was at its lowest. After 2008, it gradually grew and peaked in 2010. Although the price was lower in 2015, it was relatively stable.

In 2017, I hired 17 local diggers, and they were all very familiar with each other. All the workers are on piecework—6 yuan for each piece. The reason for the low price is the great geographical conditions and convenient transportation. In 2017, the output of the pasture was considerable—more than 20,000 pieces (more than 10 kg, 2,000 pieces for each kilogram), and the average sale price was 60,000 yuan/kg. The operating scale is relatively small. If Cordyceps from this pasture is in short supply, I will purchase the Cordyceps from the market to sell to customers. In 2017, in fact, I purchased 4 kg of Cordyceps (hay, the wholesale price of 52,000–5.4 million yuan/kg, packaged with a gift box, 30 yuan for a package) from the surrounding shops and sold to customers in Yunnan.

Source: Research group interviews, Qinfen Lane, Xining, May 2018.

Case 10: Businessman C in Jiuying Market, 'ZZH,' Male

C is a trader in the Jiuying Cordyceps market, Han nationality, from Qinghai (Huangzhong County), born in the 1960s. He has been engaged in the sale of Cordyceps for more than 20 years but doesn't sell processed products such as lozenges. The business scale of Cordyceps is about 200–300 kg per year. C mainly purchases Cordyceps from market vendors and bases instead of renting pasture and contracting Cordyceps collection.

I acted as a contractor for the first time in 2007, but later gave it up. It was mainly because of the high cost and uncertainty, and the market price demonstrated a considerable fluctuation. Therefore, I didn't dare to invest in contracting.

Now there are so many bosses, this operation is getting more and more difficult. In the past, I usually negotiated with the scattered herdsmen in the production area of Cordyceps to my advantage. Now those contractors control a large share of Cordyceps, and my negotiating status has dropped a lot.

Although the household reform began in 1983, the earliest time you could rent pastures in the market was in 2006. The cost of renting is getting higher and higher. In 2017, the rent for one mountain was 500,000 yuan/year. For example, in 2018, it increased 60% to 800,000 yuan. The reason for the price increase is mainly because the output and market price of Cordyceps increased in 2017. The digging time in Qinghai begins on May 15th and lasts for 45 days. The digging time in Nagqu, Tibet, begins on June 1—that is, the higher the altitude is, the later the work begins.

It is also difficult to manage as it not only requires a lot of time and effort but also high labor cost. Contractors have to rely on manual work for digging: 30 people for small and medium pasture, and up to 500 people for the large ones. The age of the excavators is generally between 30 and 40, both men and women, and many of them are couples.

I mainly select familiar people, not necessarily locals. Hiring familiar people is mainly to prevent the moral risk of workers because unfamiliar workers will steal the high-quality Cordyceps secretly, which is difficult to find out.

C boss cites two major standards for Cordyceps when buying from the wholesale market. One is the length and the other is the fullness (hardness), regardless of the origin of the Cordyceps. The company will also purchase Cordyceps in Nagqu because its quality is relatively high, but the price is 10,000/kg higher than that of Cordyceps in Qinghai. C's customers are mainly from Guangzhou, Shanghai, Beijing, Jiangsu, and Zhejiang. Some are commerce companies (Guangzhou Qingping Market), some are processing plants, and some are pharmaceutical companies (such as Wenzhou Chinese Medicine Company).

I once tried to sell Cordyceps on consignment to increase sales but it ended in heavy loss. I put the Cordyceps in the consignment store, where the goods could be sold. The wholesale seller would not make any payments to me in advance but would return the sales income to me according to the sales. However, the consignment store did not make an effort to sell the Cordyceps, and eventually many of them could not be sold as high-quality Cordyceps because they were left for too long. After that I strived to establish direct contact with customers and drew a line under the consignment process.

> *I did not open an online store, but use WeChat to accept orders and receive money. With the high popularity of WeChat, the information of the entire industry is too transparent. For the company, its profit margin is declining, which began in 2014. Before 2014, the average profit rate remained at 3,000 yuan/kg, but now the average profit rate has fallen to 2,000 yuan/kg, and this decline was mainly caused by the transparency brought about by the Internet.*
>
> Source: Research group interviews, Qinfen Lane, Xining, May 2018.

Conclusion

The Qinghai Cordyceps case study provides an interesting opportunity to discuss and compare how different land rights regimes can address the tragedy of the commons in the public management of natural resources and how this management has evolved in the Chinese context.

At the beginning, due to the collective nature of land rights and lack of management of Cordyceps harvesting, the Qinghai pastures reflected, at least on paper, Hardin's 'tragedy of commons': unorganized access to pastures meant anyone, both local and outsider, could dig as many Cordyceps as they wanted ('separate digging'). However, the resource was in abundance and was harvested by the locals mainly for personal consumption (Cases 9.1 and 9.2). Thus, the main reason why a CPR dilemma didn't surface was small demand and low price for Cordyceps, which disincentivized large harvesting.

The external factors, such commercial Cordyceps market growth and the rising prices per unit throughout China, but not the initial lack of institutions coordinating access to the resource, created a tragedy of commons, once the 'Cordyceps collectors' army' infiltrated these remote pastures, attracted by quick and high returns (Case 9.3).

Hardin's individual privatization of land was not applicable in the Chinese context, for institutional and ideological reasons (model D. from Table 9.1 is unapplicable). Under the rural land reform, a rural collective was serving the function of public manager, yet as it was a transitional period, the contractual rights were slowly established and transferred to new owners over an extended period of time, and the government (county, township level) in addition to local community (village) had the power to grant access to resources with the use of the so-called digging certificates. Thus, such a governance mode is more like model A rather than model B from our table. And as Ostrom warned would be the case with such a model, local social and institutional capital was destroyed: actual pasture rights owners were excluded from the decision-making process and profit-making; they were not involved in resource management, let alone the protection of Cordyceps pastures, and the resource was extracted by outsiders without traditional knowledge of how to harvest Cordyceps less intrusively, as locals would do (Case 9.3).

Once the government realized that pasture use rights holders (local herdsmen) had been denied actual rights due to the government's participation in key aspects of resource management (granting access to the resource, deciding on the numbers of digging certificates, taking their share from the profit made by selling certificates), they withdrew. Thus, the latest governance model is based on two dimensions: household and village, and has been closest to the independent grassroots governance model (C) so praised by Ostrom. Gradually, the new governance began to limit the illegal permits and certificate business (Case 9.6).

The bad news is, however, that under this new management mode, where land use rights have finally been secured and herdsmen are in the position to decide who can harvest Cordyceps, at what scale and price, from their pastures, the resource is still not managed sustainably: local herdsmen prefer to transfer the organization of labor for Cordyceps harvesting to third parties ('second, third level bosses'), who in turn are motivated to gain immediate profit; since they are not the land owners, they hire as many Cordyceps collectors as they can. The result of the new governance model is negative: overharvesting is still taking place.

Another problem is the appearance of a fake certificate industry, since locals prefer to receive revenues from smuggling in outsiders to collect Cordyceps, rather than collect it themselves (Cases 9.5, 9.6, 9.7). The introduction of government-controlled governance was partially solved not by implementing the new model of independent local governance but by the arrival of new technologies and the demand of the biggest buyers for the transparency and legality of the Cordyceps harvesting (Cases 9.8 and 9.10).

These findings contribute to the discussion of 'the institutional chicken and egg,' and revisioning neo-liberal postulates of institutional change (Ho, 2013). Our case has shown that the key driver for resource overharvesting was not just 'empty' institutions, which gradually matured, but a combination of external and internal factors; firstly the rise of external demand in Cordyceps and the low income of the rural population, which pushed high numbers of outsiders to come to the area to harvest Cordyceps, and the weakness of the newly established pasture tenure institutions to prevent overharvesting and distribute revenues among the stakeholders in a fair manner.

Overall, the Qinghai Cordyceps pasture management case study presents empirical findings that should be added to modern reflection on the CPR dilemma. While some authors have argued that excluding the farmers from the process of policy formulation and implementation was the key factor that led to Chinese grassland degradation (Wang et al., 2013), the actual expectation that local herders will manage natural resources sustainably once their use rights are secured (Thwaites et al., 1998) hasn't been realized, as there are other factors, such as preference for immediate short-term gain and lack of environmental awareness, as our case study has shown. In this scenario, the most plausible solution will be that offered by Hardin (1998) in his second theoretical update—to limit the number of users who can access the commons.

References

Awangjiancuo (2004). Significance and role of pasture household contracting system for economic social development and ecological protection in pastoral areas [in Chinese]. Proceedings of China Grass Industry Sustainable Development Strategy Forum.

Cai, B. (2014). Trading patterns of *Cordyceps sinensis* in Tibetan regions—Take 'floating' *Cordyceps sinensis* of Golog Tibetan prefecture of Qinghai as example [in Chinese]. *Qinghai Journal of Ethnology*, 1, 132–135.

Coase, R.H. (1960). The problem of social cost. *Journal of Law and Economics*, 3, 1–44.

Deininger, K., & Binswanger, H. (1999). The evolution of the world bank's land policy: Principles, experience, and future challenges. *World Bank Research Observer*, *14*(2), 247–276.

Fan, C. (2016). Economic form and cultural change of *Cordyceps sinensis* in Qinghai-Tibet area [in Chinese]. *Folklore Studies*, 1, 118–128.

Hardin, G. (1968). The tragedy of the commons. *Science* (American Association for the Advancement of Science), *162*(3859), 1243–1248.

Hardin, G. (1994). The tragedy of the unmanaged commons. *Trends in Ecology & Evolution*, *9*(5), 199.

Hardin, G. (1998). Extensions of the tragedy of the commons. *Science*, *1280*(5364), 682–683.

Ho, P. (2013). In defense of endogenous, spontaneously ordered development: Institutional functionalism and Chinese property rights. *Journal of Peasant Studies*, *40*(6), 1087–1118.

Li, A., Wu, J., Zhang, X., Xue, J., Liu, Z., Han, X., & Huang, J. (2018). China's new rural 'separating three property rights' land reform results in grassland degradation: Evidence from Inner Mongolia. *Land Use Policy*, 71, 170–182.

Ostrom, E. (1990). *Governing the commons*. Cambridge University Press.

Seligmann, L.J., & Estes, B.P. (2020). Innovations in ethnographic methods. *American Behavioral Scientist*, *64*(2), 176–197.

Theesfeld, I. (2019). The role of pseudo-commons in post-socialist countries. In B. Hudson, J. Rosenbloom and D. Cole (Eds.), *Routledge Handbook of the Study of the Commons* (pp. 345–359). Routledge.

Thwaites, R., De Lacy, T., Hong, L.Y., & Hua, L.X. (1998). Property rights, social change, and grassland degradation in Xilingol Biosphere Reserve, Inner Mongolia, China. *Science and Natural Resources*, *11*(4), 319–338.

Wang, H., Shao, J., & Cai, M. (2013). Is self-governance of the commons feasible in the PRC? A case study of pasture governance in Zhua Xixiulong township, Gansu Province. *Asia Pacific Journal of Public Administration*, *36*(3), 211–219.

Xu, M. (2016). The changes and prospects of China's agricultural land system in the 40 years of reform and opening-up [in Chinese]. *Oriental Law*, 5, 72–79.

10 Climate Change and Conservation Policy

Jiping Sheng, Jiahao Shi, Shenghang Wang, and Ksenia Gerasimova

The Qinghai-Tibetan Plateau is a vast territory with a complex topography, standing at an average of more than 5,000 meters above the sea, which makes it suitable to host *Cordyceps sinensis* in China (Shashidhar et al., 2013). Any change in climatic conditions in the Qinghai-Tibetan Plateau has a significant impact on *Cordyceps sinensis* production. Studying climate change and its impact on the Qinghai-Tibetan Plateau is scientifically useful for all aspects of *Cordyceps sinensis* production. This is also important for the implementation of climate safety. In this chapter, we will talk about the climate change conditions in the Qinghai-Tibetan Plateau, the influence on the local *Cordyceps sinensis* industry, what challenges the Cordyceps industry is facing, and how to adjust the industry in the future.

Climate Change in Qinghai-Tibetan Plateau Region

Climate change adaptation is a critical development issue (Von Borgstede et al., 2013). Experts have increasingly concluded that the Earth is becoming more susceptible to climate damage (Al-Amin et al., 2015). Climate change is evidenced by phenomena such as global warming, floods, and atmospheric events like El Nino and La Nina, which affect various aspects of the natural ecosystem. Stable climate conditions are crucial for maintaining ecosystems' health and preserving biodiversity, which in turn benefits local communities. Environments with simpler ecosystems and less biodiversity are more fragile and vulnerable to climate change. Altitude and temperature are additional factors influencing ecosystem resilience.

The Qinghai-Tibetan Plateau, the primary *Cordyceps sinensis*-producing area is one such representative area affected by climate change. Diverse ecosystems thrive in the Qinghai-Tibetan Plateau due to its varied geography and temperature (Sills et al., 2018); however, the Qinghai-Tibetan Plateau has seen a greater temperature increase in recent decades compared to other regions (Dong et al., 2012), and the continued rise (Kang et al., 2010) will further create socio-ecological challenges. The impact of the temperature rising in the Qinghai-Tibetan Plateau will be explored below.

DOI: 10.4324/9781003429753-14

The Impact of Climate Change on the Qinghai-Tibetan Plateau Region

Western China stands as one of the world's most vulnerable ecosystems and environments (Xia et al., 2021). Sparse vegetation, large temperature differences, and other factors render the local ecosystem less resilient and susceptible to external influences. In the Qinghai-Tibetan Plateau, rising temperatures will lead to the rise of the snow line (Deng et al., 2021), and the fast melting of glaciers (Moulton et al., 2021) will lead to rising water levels of local mountainous lakes, grassland degradation, and expanding desertification (Liu et al., 2017). Over the past 40 years, temperatures on the Qinghai-Tibetan Plateau have notably risen, particularly during the dry season. Rising temperatures have already disrupted the ecological cycle, as melting glaciers and receding snow lines release more water into the system. Melting glaciers and snow lines deplete water resources (Xiao et al., 2021). After initial increased transpiration, drying soils will lead to desertification threaten species, including the caterpillar fungus.

China's Response to Climate Change and Environmental Protection Policies in the Qinghai-Tibetan Plateau

Chinese central government and the local government of Qinghai Province and Tibetan regions designed a set of policies to deal with the challenges of climate change. Although the environmental conditions of the Tibetan region have improved as a result of consecutive Chinese nature conservation programs, more needs to be done.

The start of Chinese climate conservation action could be dated to 1972, when the country participated in the United Nations Conference on the Human Environment. In 1979, the first law concerning environmental conservation in China was implemented, and the process of legalization of environmental protection began. From 1979 to 1993, China continued to make progress in environmental management and protection. In March 1994, the State Council adopted 'China's Agenda 21,' elevating the overall goal of sustainable development to a national strategy (Zhang et al., 2007). On February 16, 2005, the Kyoto Protocol signed by the State's parties within the United Nations Framework Convention on Climate Change entered into force (Li, 2000). China has actively participated in multilateral environmental negotiations and participated in global environmental governance with a more open attitude and a spirit of practical cooperation. In 2008, the construction of the National Satellite Environmental Application Center began, and the successful launch of small satellites for environmental and disaster monitoring marked a new stage of development (Zhang et al., 2020). Since 2013, the state has listed the protection of 'the ecological civilization' as one of the key national development strategies and put forward higher requirements for environmental protection. In 2020, the State Council of China put forward the goal of carbon neutrality and aim to achieve it as early as possible, to cope with global climate change. In 2021, China issued the 'National Ecological Environment Protection Plan (2021–2035),' which clarified the overall goal of ecological environment protection and put forward key plans to strengthen ecosystem protection and improve environmental quality.

With an average altitude of more than 5,000 meters, Tibet has a unique natural ecology and geographical environment. The climate of Tibet changes from warm and humid to cold and arid, southeast to northwest, and the natural ecology changes from forest, scrub, meadow, and grassland to desert. The complex and diverse landforms and special ecosystem types have created great difficulties for local authorities in implementing environmental protection. In 1951, the central government sent an expedition team to assess Tibet's natural resources: land, forests, grasslands, water, and mineral resources, in order to formulate a strategy for its scientific development and utilization, thus starting the process of modern management and protection of Tibet's ecological environment (State Council, 2003). Regional comprehensive laws and regulations concerning ecological and environmental protection have been formulated in TAR, such as the 'Regulations of the Tibet Autonomous Region on Environmental Protection' (Lhasa City Ecological Bureau, 2020). In addition, the Chinese central government and the regional government of Tibet have also developed a joint vision for improvement of ecological protection and natural resource management in Tibet (Table 10.1).

Table 10.1 Environmental projects in the Qinghai-Tibetan Plateau region from 2014

Policy Title	*Issuing Authority*	*Planning Period*	*General Content*
Western Development Strategy	State Council	2014	Measures include strengthening ecological protection and promoting the development of clean energy.
Tibet Ecological Protection and Construction Plan (2015–2030)	Tibet Autonomous Region Government	2015	Measures include strengthening grassland restoration and promoting energy conservation and renewable energy utilization.
Tibet Ecological Security Barrier Construction Plan (2016–2030)	Tibet Autonomous Region Government	2016	Measures include strengthening soil and water conservation and promoting the restoration of degraded grasslands.
Western Development Ecological Compensation Policy (2017)	State Council	2017	Promotes ecological environmental protection and sustainable development in Western regions by incentivizing local governments and enterprises to take environmental protection measures.
Tibet Clean Energy Development Plan (2018–2030)	Tibet Autonomous Region Government	2018	Promotes lowering greenhouse gas emissions and the development and utilization of clean energy in the Qinghai-Tibet Plateau to reduce reliance on traditional fossil energy.

According to a report of IPCC, every region of the world is facing unprecedented changes due to climate change, from rising sea levels and frequent extreme weather events to rapidly melting sea ice. Further temperature increases will exacerbate these changes, making extreme heat, heavy rainfall, and regional droughts more frequent and severe. According to the recent data, when the average temperature increases by 1.5°C, 2°C and 4°C, the frequency of heatwaves will likely increase by 4.1, 5.6, and 9.4 times, respectively, and the intensity of heatwaves may also increase by 1.9°C, 2.6°C and 5.1°C, respectively. If global temperatures rise by more than 1.5°C, regions that rely on melting snow and ice will likely face water shortages (Lee et al., 2023). The Tibetan Plateau is the main water supply source for many major rivers and rich ecosystems in East and Southeast Asia. In Tibet the temperature is rising 3 times faster than average, and 82 Tibetan glaciers have retreated in the last 50 years (Yao et al., 2012). If the temperature increases further, the middle and lower parts of the Yangtze River, Yellow River, Lancang River, and Brahmaputra River will be affected by drought, and it will lead to the degradation and desertification of the Qinghai-Tibetan Plateau. Naturally, these adverse effects will also affect the *Cordyceps sinensis* industry.

Effects of Climate Change on the *Cordyceps sinensis* Industry and Local Households

Human activities have had a direct impact on climate change. Climate change in Tibet has in turn brought challenges to the lives of local people, and the *Cordyceps sinensis* industry, as a local industry, has also been affected to a certain extent.

Historical data suggests that climatic fluctuations in Tibet, in addition to poor governance, have created famine and political instability. In the period of the late 1200s to the mid-1400s, central Tibet was ruled by local aristocracy appointed by the Mongols. Droughts induced by climatic changes led to severe famines and further political instability (Gumble et al., 2022). The disappearance of the once flourishing Tibetan kingdom Guge in Western Tibet can also be attributed to fluctuations in temperatures; a decline in barley crops was followed by the collapse of the kingdom in the 1630s (Liang et al., 2022).

In the present period, the frequency of drought disasters and hazards such as flash floods are predicted to increase, and agricultural production may be significantly effected (Jiang et al., 2022). The onset of climate change in the Tibetan Plateau region has had a profound impact on the lives of local people. According to Wu et al. (2021), rural households in Zamtang County, located in the eastern area of the Tibetan Plateau region, have already experienced climate change: about 89% of them indicated that droughts were occurring more frequently than ever.

In parallel, to some extent, climate change, dominated by rising temperatures, has brought some benefits to local agricultural development, providing favorable climatic conditions for agricultural land reclamation. Particularly in farming and pastoralist areas at high altitudes, the warming climate has enabled households and herders to convert grasslands previously unsuitable for crop cultivation because of cold, harsh natural conditions into farmland and increase the area of cultivated land, thereby increasing their income. However, in the long term, the increased intensity of farmland reclamation and grazing would significantly diminish water conservation and carbon sequestration (Dong et al., 2020), lead to more severe soil erosion (Wang et al., 2010), reduce the quality of the soil (Gu et al., 2019), and reduce local biodiversity (Li, 2014), which may affect the sustainable development of local related industries.

In our interviews in TAR in 2018 and in similar research (Wu et al., 2021), farmers and herders indicated that they would reclaim land to increase crop cultivation acreage if they perceived climate anomalies such as the decrease in precipitation to be balancing poor yields. Cold spells also have an impact on households' herding activities; it is common for herders to expand or reduce their animal stock according to climatic conditions. Local governments also play an important role in households' response to climate change. This includes the provision of livelihood subsidies, subsistence goods, and assistance with farmers' migration to prevent the deterioration of the natural environment (Yeh et al., 2014).

The western part of China is located on the first level of the three-level terrain ladder in China, and the overall elevation is higher. A few higher plains suitable for agricultural cultivation and production do not have fertile soils. Therefore, animal husbandry has long been the main industry in the region. Under these conditions, most households adopt a self-sufficient production mode, and they rarely communicate with the outside world, so local economic integration with outside regions has been very slow. However, the harsh natural conditions in this area also provide natural advantages for the production of *Cordyceps sinensis*: the sparse local population and low-temperature environment of the higher altitude provide good conditions for *Cordyceps sinensis* growth. Glacial meltwater provides the water that *Cordyceps sinensis* needs. According to archaeological results, the history of picking *Cordyceps sinensis* in China probably dates back 2,000 years.

The growth mechanism and development process of wild *Cordyceps sinensis* still remain a mystery. That's why there are still many difficulties in artificially cultivating it, as it is extremely difficult to simulate the extremely cold temperatures and low oxygen of the natural environment, and a high ultraviolet irradiation intensity is needed in the laboratory. Therefore, the main source of *Cordyceps sinensis* is still from nature.

The northwestern region of China is a vast territory with complex topography, and climate warming varies from place to place. The areas that have experienced the most obvious increase in temperature are in the northwestern part of Xinjiang, the Junggar Basin, and the Turpan Basin. Temperatures have also risen in the eastern part of Xinjiang and elsewhere in the Qinghai Plateau (Li et al., 2012). It has produced several negative trends, such as grassland degradation and melting glaciers.

Grassland Degradation and Cordyceps sinensis

In the past 40 years, the grassland system in the northwest of Tibet has undergone profound changes, and the forage season lasts longer. The central and eastern parts of northwestern Tibet have experienced rising temperatures, which was initially conducive to grassland productivity. However, as the average temperature in the western region continued to increase, droughts became more frequent, which is naturally extremely detrimental to grass growth (Lu et al., 2017). From the perspective of climate change trends in the Qinghai-Tibetan Plateau, grassland degradation stands out as worryingly serious in northwestern Tibet (Zhou et al., 2023). With grassland degradation comes problems such as insect infestation and soil erosion.

The grassland environment is the main ecosystem of *Cordyceps sinensis*, which has specific environmental requirements and does not grow in ordinary meadows. In terms of temperature, *Cordyceps sinensis* needs the temperature to be low at first and then high. Although low temperatures mean the slow growth of the fungus' mycelium, it increases the amount of time it has to accumulate nutrients and reduces the survival rate of other fungi, which can significantly improve final product quality. It is worth noting that *Cordyceps sinensis* is resistant to low temperatures, but temperatures higher than 32°C may kill it. Therefore, although climate warming in northwestern China promotes cultivation of wintering crops, it has a negative effect on the Cordyceps industry—the time of nutrient accumulation in fungi is shortened and the quality of Cordyceps is reduced. *Cordyceps sinensis* requires soil to have a certain amount of water content and good water permeability to avoid excessive water affecting its growth. In the eastern part of the northwestern region, an increase in droughts means soil water evaporation is intensified (Cao et al., 2021). Climate warming has also contributed to the amount of degraded and desertified grasslands in pastoral areas, which decreases grassland production capacity, including *Cordyceps sinensis* production. In addition, the warming of Tibetan winters has not only increased the evaporation of soil moisture in winter pastures but also affected the safe wintering of *Cordyceps sinensis*. It reduces the mortality of wintering pests and diseases and increases the difficulty of preventing infection in grass plants (Zhao et al., 2014).

Melting Glaciers and Cordyceps sinensis

The warming of the Qinghai-Tibetan Plateau may cause glacier retreat, erratic changes in snow cover, and the melting of permafrost, all of which have an impact that extends well beyond the Qinghai-Tibetan Plateau itself, affecting the Earth's atmospheric circulation and the water supply for billions of people living downstream (Qiu, 2008). As mentioned above, the retreat of snow lines and melting glaciers is also happening in northwest China. It has a profound impact on the ecosystem, the lives of local people, and the cultivation of *Cordyceps sinensis*. In summer, as temperatures rise across the region, glacier melting increases, putting greater pressure on water and soil conservation in the local middle and lower river basins.

Some argue that human industrial activity causes the phenomenon; black soot rain, also called black carbon, is 50% responsible for the ice melting (Buckley, 2019). An increased flow of snow and ice melt water can exacerbate erosion along the river, which will bring changes to the environment in which *Cordyceps sinensis* grows. Cordyceps have a long growth cycle and have high requirements for the growing environment, such as good quality soil and the right temperature and water content. Habitat changes caused by glacier melting will seriously affect the growth of Cordyceps and further affect harvesting and the industry as a whole.

Due to gravitational potential energy, glacial meltwater can easily cause flooding, which will cause serious damage to downstream residential and agricultural areas. At the same time, it will also bring secondary disasters such as diseases or pest infestation, which can occur after floods and pose a great threat to the lives of residents and the development of local industries.

Cordyceps sinensis Industry and Climatic Threats

Due to people's pursuit of the magical effects of *Cordyceps sinensis*, the Qinghai-Tibetan Plateau, as the main production area of *Cordyceps sinensis*, makes up quite a big part of the industry, and therefore local households are closely linked to this industry. Some threats to the industry associated with changes in climate and environment have emerged.

Due to the influence of global climate change, the snow line of the Qinghai-Tibetan Plateau has risen, and the distribution pattern of *Cordyceps sinensis* in the alpine meadow of the Qinghai-Tibetan Plateau has also changed significantly. For example, 30 years ago, the core distribution zone of *Cordyceps sinensis* was between 3,800 and 4,500 meters and has now risen to between 4,400 and 4,700 meters (Hu, 2017). In the area of 4,200 to 4,500 meters above sea level, the fungi population is decreasing year by year. Some areas have not produced *Cordyceps sinensis* for many years, as there is no sign of host insects. Under climate change, the prediction about the future of the habitat of Cordyceps is even more pessimistic (Yan et al., 2017). Under the condition of RCP2.6 (Representative Concentration Pathways with the lowest GHG emissions), the potential net habitat loss would be 37% for both years 2050 and 2070; under the condition of RCP8.5 (Representative Concentration Pathways with the highest GHG emissions), the potential net loss would be 37% and 39% for the years 2050 and 2070, respectively (Table 10.2).

More importantly, driven by huge profits, more and more people have rushed to the Qinghai-Tibetan Plateau to frantically collect this species, resulting in a sharp decline in its population. Our observations and interviews suggest that the area of *Cordyceps sinensis* harvesting in the Qinghai-Tibetan Plateau has expanded by 5 to 6 times in the last 10–15 years, but the overall production volume is the same as

Table 10.2 Predicted habitat loss of *Cordyceps sinensis* in the Tibetan region

Years	*Scenarios*	*Habitat loss*
2050	RCP2.6	37%
	RCP8.5	36%
2070	RCP2.6	37%
	RCP8.5	39%

in the past years, or has even reduced by about 50 to 80 tons. In other words, the yield per unit area has fallen. Even in some of the places where *Cordyceps sinensis* grew the most densely, the number collected by scientists and the host insects found were very few. Further, the development of the *Cordyceps sinensis* industry has not only reduced the number of this species but also had a negative impact on the surrounding ecological system.

The impact of the *Cordyceps sinensis* industry on ecological systems is revealed by the degradation of the surrounding ecosystem. In the transition between spring and summer, a larger number of people come to Tibet and Qinghai regions to dig *Cordyceps sinensis*. For the Qinghai region, according to the relative data, it can attract more than 200,000 people to come here for *Cordyceps sinensis* harvesting in just one day. The rapid increase of human activities in a short period has undoubtedly caused serious damage to the already fragile local ecological environment. Due to the weak resilience of the ecosystem, it continues to degrade. Trampling on the meadow is likely to lead to the destruction of grassland plants, thus destroying their regenerational ability. *Cordyceps sinensis* excavators usually choose to live in the local area for a while. During this period, the harvesters will carry out activities that are of great risk to the meadow, such as making fires. Also, a large amount of domestic garbage is abandoned in the meadow after they have left. In addition, because *Cordyceps sinensis* is rare, it is often grown together with ordinary grass strains. After removing fungi from the soil, it is difficult for most excavators to fill the missing sod, thus leaving a series of pits in the meadow. These pits cannot support the grassland anymore, which can lead to lead to serious desertification.

The Qinghai-Tibetan Plateau is known as the 'water tower of Asia,' and the core distribution area of *Cordyceps sinensis* lies in alpine meadows at the headwaters of the Yangtze River, Yellow River, Lancang River, and so on, thus playing an important ecological role in Asia. With the degradation of grassland brought on by the overharvesting of *Cordyceps sinensis*, the soil along the river becomes loose due to the reduction of vegetation, and so the rivers carry away more soil, which seriously affects the water quality in the middle and lower reaches.

Not only the ecological environment but also local social and economic aspects are affected by the *Cordyceps sinensis* industry. In TAR alone, 280 towns and villages are within the boundaries of Cordyceps habitats, and in our estimations, it might be that a few million people in Tibet and Qinghai rely on harvesting Cordyceps. Many households say that if it were not for *Cordyceps sinensis*, they would not be able to pay for everything they need, such as building and amending houses and buying daily necessities. *Cordyceps sinensis* also provides possibilities for local households to develop their businesses; the high profits made from *Cordyceps sinensis* can be used to buy agricultural inputs, such as fodder, pesticides, and fertilizers (Wu, 2008). Some households in the interviews we conducted in 2018 also said that the income from 'digging half a kilogram of *Cordyceps sinensis* is equivalent to their income from three years of working outside.'

The negative impact of the *Cordyceps sinensis* industry on society should not be underestimated. First of all, the future of the industry is highly uncertain. Since the cultivation method of artificial *Cordyceps sinensis* has not yet been commercialized, local households can still benefit from high prices for the raw Cordyceps they harvest in the natural environment. However, the availability of *Cordyceps sinensis* has been extremely reduced due to its low yield, long growth cycle, and frequent and excessive human exploitation. As mentioned above, the annual volume collected each year is decreasing. The destruction of the ecological environment brought on by over-exploitation has brought great challenges to the future development of the *Cordyceps sinensis* industry. However, increasingly local households are attracted by the high profits from collecting *Cordyceps sinensis* and are quitting their original industries, such as cultivating highland barley and animal husbandry (Danjiu, 2019; Yu, 2022). It is foreseeable that with the further deterioration of the *Cordyceps sinensis* industry, the amount of employment that can be accommodated will continue to decrease, and many households may no longer be able to profit from the industry.

There is also the issue of climatic fluctuation, and local Tibetan communities have developed several livelihood adaptation strategies, such as mobility, storage (seedbanks, additional fodder and reserve livestock), common pooling (sharing pastures and water wells), barter exchange, and pasture sublease (Wang et al., 2016). In addition, more recent strategies include livelihood diversification and collecting and trading Cordyceps. As shown previously, with the arrival of the *Cordyceps sinensis* industry, the living standards of local rural communities rapidly increased. Former cultural Buddhist practices of restrain and modesty have been replaced with sometimes irrational consumption after sudden increases in wealth from trading caterpillar fungi. It is common for households to spend the money they earn unrestrainedly rather than re-invest it into reproduction.

Cordyceps sinensis harvesting presents a very high security risk to the local ecological environment and to economic and social ecology. Therefore, regulating the development of the industry in Qinghai-Tibet and promoting the sustainable development of the region is one of the problems to be solved urgently.

Existing Measures to Mitigate Adverse Effects and Promote the Sustainable Development of the *Cordyceps sinensis* Industry

Local households and the government have combined their efforts to maintain a balance between production of Cordyceps and the stability of the local ecological and social environment. For local households, the most recent climate change adaptation strategy consists mainly of relocating harvesting and searching for alternative activities. While transferring collection sites can temporarily boost volumes of collected fungi, harvesting unsustainably and without implementing ecological restoration actions will inevitably result in ecological damage in these new areas. To avoid harming the worm, collectors often remove the turf surrounding it. Few individuals backfill the turf and soil and in the long run contribute to the degradation of grasslands (Danjiu, 2019).

Some households engaged in the industry have difficulty making a living from *Cordyceps sinensis* alone, and thus have already had to revert to local traditional industries, such as grazing and agriculture (Interview with Tibetan farmers, Namtso Lake, May 2018). A similar trend has been reported for Bhutan (see Chapter 6).

To ensure the sustainability of the *Cordyceps sinensis* industry and promote ecological protection, government departments in the Tibetan Plateau region have established regulations including Presidential Decree No. 70 of the People's Government of the Tibet Autonomous Region. Additionally, regional governments have intensified *Cordyceps sinensis* resource management and enacted various policies including implementing fees for resource management and issuing collection certificates. During excavation periods, authorities visit each pasture to control the number of harvesters present at the site. Simultaneously, these government entities are collaborating with scientific research institutions to explore artificial cultivation methods for *Cordyceps sinensis*, aiming to mitigate ecosystem damage from natural digging.

The Future of *Cordyceps sinensis* in the Context of Climatic Challenges

Despite the many challenges, there is no denying that the *Cordyceps sinensis* industry has brought unimaginable prosperity to the local area. In the context of further climate change in the future, how should local communities promote further sustainable development of the *Cordyceps sinensis* industry?

A survey we did among households involved in the *Cordyceps sinensis* industry provided further evidence of this economic prosperity. 'Digging *Cordyceps sinensis* is the most important source of income for the family, and we should not miss any *Cordyceps sinensis* season,' said Dava, a native of the Qinghai-Tibetan Plateau. Dava has been digging for more than 30 years. According to him, most of the living expenses of the families in his village depend on the income from selling *Cordyceps sinensis*.

In recent years, with the improvement of average Chinese people's living standards, health issues have become a new priority, and *Cordyceps sinensis* has been perceived as 'a key to the door of health for wealthy people.' But the limited resources of *Cordyceps sinensis* means market demand cannot be met, causing prices to soar year after year. 'For several years, some good *Cordyceps sinensis* has been sold for more than 200,000 CNY per kilogram,' said a government official from Nagqu, which is famous for high-quality *Cordyceps sinensis* production within the Qinghai-Tibetan Plateau region. In Baja Village, a village that once relied on animal husbandry, the local ecological environment is very suitable for the growth of *Cordyceps sinensis*, so the industry has developed rapidly in the area. In the past, when local people lived on animal husbandry, the annual income of villagers could only meet the basic needs of their families. After entering the *Cordyceps sinensis* industry, in 2013, the average income to be earned in the village exceeded 180,000 CNY, and the income of the largest household was more than 600,000 CNY because 14 family members dug *Cordyceps sinensis* together. Lunzhu, the head of Baga village, has nine sons and three daughters, all of whom are also involved in the *Cordyceps sinensis* digging business, and their family income exceeded 400,000 CNY in 2014 (Li, 2014). The temptation of making high profits from the *Cordyceps sinensis* industry will attract more visitors to invest in digging work. At the same time, some people have set up businesses and production lines for *Cordyceps sinensis* in the production area. The establishment of relevant enterprises has promoted the improvement of the industry chain and brought higher income to the local people, which also provides better opportunities for the development of the industry.

It is undeniable that the industry is also facing a dual ecological and social threat. In terms of nature, with the development of global warming, the high altitude environment where *Cordyceps sinensis* grows will gradually be more seriously affected. Water shortages, grassland degradation and other disasters will seriously affect the quality and yield of *Cordyceps sinensis*. In terms of society, with households dependent on profits brought by *Cordyceps sinensis*, uncontrolled and unplanned collection has reduced the natural supply. With rising temperatures, the altitude of *Cordyceps sinensis* habitats has been rising. This further exacerbates the risks faced by Cordyceps collectors. A native mentioned that collectors in the sparsely populated wilderness may encounter snow leopards and brown bears as well as bad weather such as snow storms, avalanches and other threats to life. With the decline in production and the deterioration of the natural environment, the local government has tried to restrict how much *Cordyceps sinensis* is collected by issuing digging permits, but this policy has led to violent conflicts among people in different regions for the right to harvest it.

In terms of future policy development, the local government's participation is essential for promoting the healthy development of the industry. At present, the local government has standardized collection, acquisition, processing, and other elements of the *Cordyceps sinensis* industry and adopted a licensing system. Carrying out harvesting activities without a license is considered illegal and will be subject to relevant penalties.

Deeply exploring the development potential of the *Cordyceps sinensis* industry is a key focus. For example, combining the *Cordyceps sinensis* industry with cultural and eco-tourism in the Qinghai-Tibetan Plateau could play a role in diverting Cordyceps collectors to other opportunities in their local area.

In terms of the preservation of traditional industries, animal husbandry, as a means of livelihood for people in Tibet, has important historical and cultural value. To promote the conservation of traditional animal husbandry and to enable local people to make a good living outside the Cordyceps industry, local government can assist local herders through subsidies and other payments. At present, the central government provides comprehensive subsidies to households in all Tibetan areas, including those covering food, accommodation, medical care, and old-age insurance (TAR Development and Reform Commission, 2018). The government also provides interest-free loans to herders if they need money to start a new business. At the same time, local communities are searching for ways to adapt to new climatic, ecological, and economic conditions.

References

Al-Amin, A.Q., Rasiah, R., & Chenayah, S. (2015). Prioritizing climate change mitigation: An assessment using Malaysia to reduce carbon emissions in future. *Environmental Science & Policy*, *50*, 24–33. https://doi.org/https://doi.org/10.1016/j.envsci.2015.02.002

Buckley, M. (2019, July 19). Climate emergency in Tibet. *The Ecologist*. https://theecologist.org/2019/jul/19/climate-emergency-tibet

Cao, S., He, Y., Zhang, L., Chen, Y., Yang, W., Yao, S., & Sun, Q. (2021). Spatiotemporal characteristics of drought and its impact on vegetation in the vegetation region of Northwest China. *Ecological Indicators*, *133*, 108420. https://doi.org/https://doi.org/10.1016/j.ecolind.2021.108420

Danjiu, L. (2019). Negative effects of *Cordyceps sinensis* mining and development countermeasures. *Modern Agricultural Science and Technology*, 13, 91–92.

Deng, G., Tang, Z., Hu, G., Wang, J., Sang, G., & Li, J. (2021). Spatiotemporal dynamics of snowline altitude and their responses to climate change in the Tienshan Mountains, Central Asia, during 2001–2019. *Sustainability*, *13*(7).

Dong, M., Jiang, Y., Zheng, C., & Zhang, D. (2012). Trends in the thermal growing season throughout the Tibetan Plateau during 1960–2009. *Agricultural and Forest Meteorology*, *166–167*, 201–206. https://doi.org/https://doi.org/10.1016/j.agrformet.2012.07.013

Dong, S., Shang, Z., Gao, J., & Boone, R.B. (2020). Enhancing sustainability of grassland ecosystems through ecological restoration and grazing management in an era of climate change on Qinghai-Tibetan Plateau. *Agriculture, Ecosystems & Environment*, *287*, 106684.

Gu, X., Zhou, X., Bu, X., Xue, M., Jiang, L., Wang, S., & Wang, G. (2019). Soil extractable organic C and N contents, methanotrophic activity under warming and degradation in a Tibetan alpine meadow. *Agriculture, Ecosystems & Environment*, *278*, 6–14.

Gumble, R., Powers, J., & Hackett, P. (2022). Central Tibetan famines 1280–1400: When premodern climate change and bad governance starved Tibet. *Bulletin of SOAS*, *85*(2), 215–233.

Hu, M. (2017). The fate changed by Cordyceps. *Chinese Journal of Science*.

Jiang, T., Su, X., Singh, V.P., & Zhang, G. (2022). Spatio-temporal pattern of ecological droughts and their impacts on health of vegetation in Northwestern China. *Journal of Environmental Management*, *305*, 114356. https://doi.org/https://doi.org/10.1016/j.jenvman.2021.114356

Kang, S., Xu, Y., You, Q., Flügel, W.-A., Pepin, N., & Yao, T. (2010). Review of climate and cryospheric change in the Tibetan Plateau. *Environmental Research Letters*, *5*(1), 015101. https://doi.org/10.1088/1748-9326/5/1/015101

Lee, H., Calvin, K., Dasgupta, D., Krinner, G., Mukherji, A., Thorne, P., & Barret, K. (2023). *IPCC, 2023: Climate change 2023: Synthesis report, summary for policymakers. Contribution of working groups I, II and III to the sixth assessment report of the intergovernmental panel on climate change* [Core Writing Team, H. Lee and J. Romero (Eds.)]. IPCC, Geneva, Switzerland.

Lhasa City Ecological Bureau (2020). *Regulations of the Tibet Autonomous Region on environmental protection*. Lhasa.

Li, B., Chen, Y., & Shi, X. (2012). Why does the temperature rise faster in the arid region of northwest China? *Journal of Geophysical Research: Atmospheres*, *117*(D16).

Li, J. (2014). Villagers in Tibet earned more than 60 thousand yuan from digging cordyceps. *Xinhuanet*.

Li, Y. (2000). The costs of implementing the Kyoto Protocol and its implications to China. *International Review for Environmental Strategies*, *1*(1), 159–174.

Liang, J., Guo, Y., Richter, N., Xie, H., Vachula, R., & Luien, R.L. (2022). Calibration and application of branched GDGTs to Tibetan lake sediments: The influence of temperature on the fall of the Guge Kingdom in Western Tibet, China. *Paleoceanography and Paleoclimatology*, 37, 1–23. https://doi.org/10.1029/2021PA004393

Liu, Y., Liu, G., Xiong, Z., & Liu, W. (2017). Response of greenhouse gas emissions from three types of wetland soils to simulated temperature change on the Qinghai-Tibetan Plateau. *Atmospheric Environment*, *171*, 17–24. https://doi.org/https://doi.org/10.1016/j.atmosenv.2017.10.005

Lu, X., Kelsey, K.C., Yan, Y., Sun, J., Wang, X., Chang, G., & Neff, J.C. (2017). Effects of grazing on ecosystem structure and function of alpine grasslands in Qinghai-Tibetan Plateau: A synthesis. *Ecosphere*, 8(1), 1656.

Moulton, H., Carey, M., Huggel, C., & Motschmann, A. (2021). Narratives of ice loss: New approaches to shrinking glaciers and climate change adaptation. *Geoforum*, *125*, 47–56. https://doi.org/https://doi.org/10.1016/j.geoforum.2021.06.011

Qiu, J. (2008). China: The third pole. *Nature*, *454*, 393. https://link.gale.com/apps/doc/A694264283/AONE?u=anon~a46d3e49&sid=googleScholar&xid=92e0dcdd

Shashidhar, M.G., Giridhar, P., Udaya Sankar, K., & Manohar, B. (2013). Bioactive principles from *Cordyceps sinensis*: A potent food supplement—A review. *Journal of Functional Foods*, *5*(3), 1013–1030. https://doi.org/10.1016/j.jff.2013.04.018

Sills, J., Liu, J., Milne, R.I., Cadotte, M.W., Wu, Z.-Y., Provan, J., & Li, D.-Z. (2018). Protect third pole's fragile ecosystem. *Science*, *362*(6421), 1368–1368. https://doi.org/10.1126/science.aaw0443

State Council (2003). *Ecological Construction and environmental protection in Tibet*. Beijing.

TAR Development and Reform Commission (2018). *Agricultural and animal husbandry development plan for the 13*th *five year plan period for Tibet Autonomous Region*. Lhasa.

Von Borgstede, C., Andersson, M., & Johnsson, F. (2013). Public attitudes to climate change and carbon mitigation—Implications for energy-associated behaviours. *Energy Policy*, *57*, 182–193. https://doi.org/https://doi.org/10.1016/j.enpol.2013.01.051

Wang, J., Wang, Y., Li, S., & Qin, D. (2016). Climate adaptation, institutional change, and sustainable livelihoods of herder communities in northern Tibet. *Ecology and Society*, *21*(1), 5. http//ds.doi.org/10.5751/1:S-08170-210105

Wang, Y. B., Wang, G.X., Wu, Q.B., Niu, F.J., & Chen, H.Y. (2010). The impact of vegetation degeneration on hydrology features of alpine soil. *Journal of Glaciology and Geocryology*, *32*(5), 989–998.

Wu, J. (2008). *Explore the code of Cordyceps breeding*. Xinhuanet. https://m.xinhuanet.com/yn/218-09/17/c_37473933.htm

Wu, S., Yan, J., Yang, L., Cheng, X., & Wu, Y. (2021). Farmers and herders reclaim cropland to adapt to climate change in the eastern Tibetan Plateau: A case study in Zamtang County, China. *Climatic Change*, *165*(3–4), 69.

Xia, M., Jia, K., Zhao, W., Liu, S., Wei, X., & Wang, B. (2021). Spatio-temporal changes of ecological vulnerability across the Qinghai-Tibetan Plateau. *Ecological Indicators*, *123*. https://doi.org/10.1016/j.ecolind.2020.107274

Xiao, Y., Xiong, Q., Liang, P., & Xiao, Q. (2021). Potential risk to water resources under eco-restoration policy and global change in the Tibetan Plateau. *Environmental Research Letters*, *16*(9), 094004. https://doi.org/10.1088/1748-9326/ac1819

Yan, Y., Li, Y., Wang, W.-J., He, J.-S., Yang, R.-H., Wu, H.-J., Wang, X.-J., Jiao, L., Tang, Z., & Yao, Y.-J. (2017). Range shifts in response to climate change of Ophio*Cordyceps sinensis*, a fungus endemic to the Tibetan Plateau. *Biological Conservation*, *206*, 143–150. https://doi.org/10.1016/j.biocon.2016.12.023

Yao, T., Thompson, L., Yang, W., Yu, W., Gao, Y., Guo, X., Yang, X., Duan, K., Zhao, H., Xu, B., Pu, J., Lu, A., Xiang, Y., Kattel, D.R., & Joswiak, D. (2012, July 15). Different glacier status with atmospheric circulations in Tibetan Plateau and surroundings. *Nature Climate Change*. Letters. doi: 10.1038/NCLIMATE1580

Yeh, E.T., Nyima, Y., Hopping, K.A., & Klein, J.A. (2014). Tibetan pastoralists' vulnerability to climate change: A political ecology analysis of snowstorm coping capacity. *Human Ecology*, *42*, 61–74.

Yu, X. (2022). Current development of Yushu Cordyceps industry: New problems and suggestions for countermeasures. *Yushunews*. https://wm.yushunews.com/system/2022/08/03/013613223.html

Zhang, K., Wen, Z., & Peng, L. (2007). Environmental policies in China: Evolvement, features and evaluation. *China Population, Resources and Environment*, *17*(2), 1–7. https://doi.org/https://doi.org/10.1016/S1872-583X(07)60006-0

Zhang, X., Wang, F., Wang, W., Huang, F., Chen, B., Gao, L., & Li, Z. (2020). The development and application of satellite remote sensing for atmospheric compositions in China. *Atmospheric Research*, *245*, 105056. https://doi.org/https://doi.org/10.1016/j.atmosres.2020.105056

Zhao, H.-Y., et al. (2014). Climate change impacts and adaptation strategies in Northwest China. *Advances in Climate Change Research*, *5*(1), 7–16. https://doi.org/https://doi.org/10.3724/SP.J.1248.2014.007

Zhou, H., Yang, X., Zhou, C., Shao, X., Shi, Z., Li, H., & Hu, X. (2023). Alpine grassland degradation and its restoration in the Qinghai–Tibet Plateau. *Grasses*, *2*(1), 31–46.

11 Poverty Alleviation and Wild *Cordyceps sinensis* Collection

Jiping Sheng, Jinshuo Zhang, and Ksenia Gerasimova

Introduction

The main production areas of *Cordyceps sinensis*—Yushu and Golog in Qinghai Province, Nagqu and Qamdo in Tibet, Ganzi and Aba in Sichuan, and Gannan in Gansu Province—have been traditionally populated by Tibetans. Tibetan areas of the four provinces (Qinghai, TAR, Sichuan, and Gansu) are known as 'the national contiguous areas' with a particularly high level of poverty. Due to its location 'on the roof of the world,' in the hinterland of the Qinghai-Tibet Plateau, and extremely difficult natural conditions, including cold climate, barren soils, and high altitudes, the population density of the region is low. Among the 14 Chinese poorer areas, the four mentioned provinces have been the poorest in the country (Galden & Li, 2021). The problem of poverty alleviation in the region has become the most prominent, and it has been an arduous task to achieve the goal of building an evenly well-off society in China. In this chapter, we explore the role of harvesting Cordyceps in the poverty alleviation policy of China. We have used socio-anthropological data received from fieldwork conducted in Qinghai Province, Nagqu and Ganzi, and Linzhi, Lhasa in Tibet.

Previous Efforts to Alleviate Poverty via Chinese Policy in the West

The Chinese government made efforts to address the socio-economic disparity of rural and urban areas during the Great Leap Forward (1958–1960). The Second Five Year Plan (1958–1962) first included resettlement programs in poorer areas, such as Inner Mongolia (Robin & Hall, 2009). The next efforts to alleviate poverty, including in the Tibetan highlands, came with the economic reforms in the late 1970s. In 1994 the Working Panel on Tibet introduced housing projects for needy farmers and herdsmen. In 1999 a new regional development strategy 'Western Great Development' (Xibudakaifa) consisting of three stages was introduced by the Chinese government. It focused on infrastructure development, environmental preservation, industrial restructuring, investment and upgrading skills (Jeong, 2015). In 2002 Hu Jintao proclaimed a new target of achieving a 'harmonious society' to reduce the gap between urban and rural living standards, which was further supported by the concept of 'the new socialist rural life' at the Sixteen Congress

DOI: 10.4324/9781003429753-15

of the Communist Party (Robin & Hall, 2009). Although some scholars considered this program 'a civilising project' of the Chinese state that symbolized contradictions of 'center' vs 'periphery,' even they admitted that these poverty alleviation efforts have reduced the number of those living in poverty in the region from over a million to less than half a million people (Jeong, 2015). This 'Go West' policy created a new network of railroads that connected Qinghai to Tibet and transformed Tibet into a trading hub in the Himalayas (Mathou, 2005).

Part of the regional development of China's West was a policy of resettling nomads, which was not exclusive to the Tibetan region, as we earlier explained. In Qinghai Province over 700,000 nomads were relocated into 'New Happiness Villages,' in order to alleviate rural poverty and improve environmental conservation, since nomadic livestock and land management practices were believed to be 'inefficient.' These relocations received a mixed reaction: on one hand, nomads and their children might have lost their place-knowledge that previously pushed them to make 'socially consonant and ecologically adaptive decisions,' on the other, they received access to adequate educational and medical services (Bauer, 2015). Interestingly, the field visits to former nomads in their relocated homes in Qinghai Province revealed that they had developed an alternative income-generating activity—collecting Cordyceps (Bauer, 2015).

Since the 1990s, *Cordyceps sinensis* has become such a popular medicinal material; demand for the fungus has increased domestically and internationally (Pradhan et al., 2020; Galden & Li, 2021). As a result, prices for raw Cordyceps started to rise rapidly. In November 2023, the price reached 121.14 RMB per gram, and processed Cordyceps products could cost 300–500 RMB per gram, compared with the 2017 price of 350 yuan per gram (Fourth National Survey on Chinese Materia Medical Resources, 2017). This explains how collecting Cordyceps is intertwined with China's poverty eradication process, with many poor counties and villages relying on Cordyceps to earn their income. But there are other reasons too.

Thus, further enquiry into existing patterns of harvesting fungi, which are an example of a non-timber forest products (NTFP), defined as 'biological resources of plant and animal origin, harvested from natural forests, manmade plantations, wooded land, farmlands, trees outside forests and or domesticated' (FAO, 1999), can reveal the strategies available to the rural poor to improve their income. The Tibetan case of harvesting Cordyceps also illustrates serious transformations in the Himalayan ecological and socio-economic systems. Collecting these precious fungi has become the key income-generating activity of rural residents of the Qinghai - Tibetan region, and this study on Cordyceps presents important insights for future policy amendments.

The Transformation of the Tibetan Regional Economy

For centuries Tibetan society has been under the remarkable influence of its religion—Buddhism. In the past, nomads made up roughly half of the population, with tradesmen and agriculturalists, and priests, monks and nuns, making up the other half (Chapela, 1992). Such division of labor came from food production capacities; most arable land in the past was used for production of Tibetan barley (tsampa) and wheat, two staple foods in the region (Wiley, 1986). Animal husbandry provided milk and meat, essential protein in the harsh climate, and wool and animal skins were key trading items. Land was divided between aristocratic families, government, monasteries and small farming communities. In the past, Tibetan farming communities (takser) were 'self-sufficient,' and a system of seed banks functioned as a reserve in case of poor harvests. The religious influence and harsh climate did not promote consumerism or entrepreneurial activities (Chapela, 1992).

The collectivization in Tibet was aimed at bringing 'the same amenities to the mountain dwellers in remote locations as were available to members of collective farms and people's communes in other rural areas' (Kreutzmann, 2011, p. 208). Peter Ho has demonstrated that the collectivization policy actually aimed to secure food security for Tibetan farmers (Ho, 2003). A contested element, resettlement was also aiming to provide nomads in remote high altitude locations access to social services and infrastructure, although relocating to new townships came at the expense of longer travel to summer pastures (Kreutzmann, 2011).

Privatization, and its introduction of the 'Household Responsibility System,' came to TAR after 1980, and the process of dissolving communes was welcomed by the Tibetan pastoralists. However, the majority of rural herdsmen did not have access to urban markets for agricultural commodities, as these were controlled by Hui merchants (Bauer, 2005). Wang (2014) mentions other missed opportunities in the areas of education and skills training; lower levels of innovation meant an influx of skilled laborers from outside the region.

Thus, at the beginning of the new economic policy of China, Tibetan farmers in TAR and Qinghai were poorer than average Chinese farmers and the gap was widening. In 1985 the average Chinese farmer's income was ¥ 398 ($64) against a ¥ 353 ($54) income in TAR. In 1996, these figures were ¥ 2,090 ($337) and ¥ 975 ($157) (Wang, 2014). By 2007 rural TAR had moved from a subsistence model into a new mixed economy, as 70% of total incomes for rural families came from non-farm activity (Goldstein et al., 2008). In 1999 revenues from the collection and trade in Cordyceps amounted to 6.55% of TAR GDP, while in 2007 it was 15.7%. In the key Cordyceps-producing areas, harvesting Cordyceps contributed to 80% of all household income, particularly in Nagqu, Qamdo and Nyingchi (Mu et al., 2011). According to Luorong and Dawa (2006), 750,000 rural residents in TAR depended on the revenues from collecting *Cordyceps sinensis*—including 450,000 people in Qamdo and 200,000 people in Nagqu, the two key Cordyceps-producing areas in the region.

Current Economic Development in the Main Production Areas of *Cordyceps sinensis*

Under numerous Chinese state programs, including generous subsidies, Tibetan areas in the four above-mentioned provinces achieved rapid economic growth, averaging 15.4% per year during the '12th Five-Year Plan period' (2011–2015). To

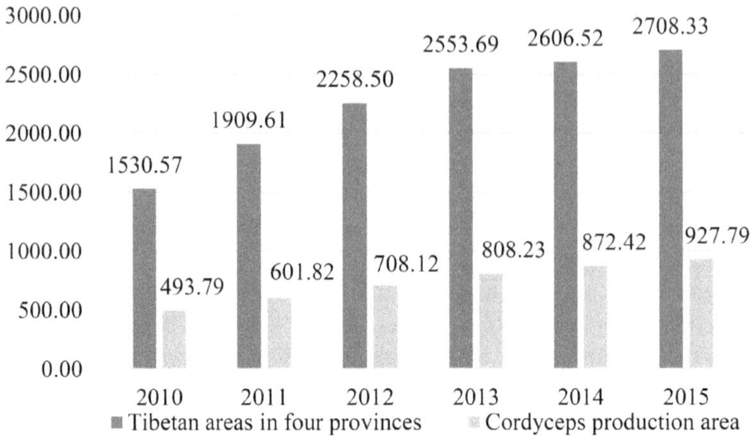

Figure 11.1 The Gross Regional Product of the four provinces against the Gross Domestic Product of the *Cordyceps sinensis* production areas (Unit: 100 million yuan)

Compiled using the Statistical Yearbooks 2011–2016 of Gansu, Tibet, Qinghai, and Sichuan.

compare, in this period GDP of the Cordyceps-producing areas in these provinces achieved an average annual growth rate of 17.6% (see Figure 11.1).

By 2012, at the beginning of the new poverty alleviation campaign, there were still 590,000 registered poor people in Tibet, many of whom lived in high altitude areas and with poor infrastructure. A part of the poverty eradication campaign was the relocation of farmers to new townships. Chinese media widely reported how Tibetan farmers overcame their initial frustration about relocation and appreciated 'the resettlement of the new village with convenient transportation, complete energy facilities, and good working and living conditions, so that the villagers have more confidence to chase new dreams' (Gama, 2020).

During the '13th Five-Year Plan period' (2016–2020), the Tibet Autonomous Region documented that all 628,000 poor people were moved out of poverty, and the per capita net income of the poor population had increased from 1,499 yuan in 2015 to 9,328 yuan in 2019 (ibid.). The GDP growth rate in 2023 reached 9.8%, ranking first in the country. By the end of 2022, the total population of the main production areas of the four provinces reached 143.80 million, of which the Tibetan population accounted for more than 95%, and other ethnic groups included Hui, Mongolian and Han.[1]

Indeed, harvesting *Cordyceps sinensis* is the main economic source for the people in the Tibetan areas of China, but it is not by choice; other activities are becoming more difficult to maintain. Not one family can rely on farming activities for its total income; everyone has sought non-farm employment to support their families (Wang, 2014).

More than 80% of farmers' and herdsmen's families in the main production areas rely on selling Cordyceps, which accounts for 50%–80% of total income. In TAR the coverage of Cordyceps-producing pastures is 6,500 million mu, and in best years, collectors could harvest 40–50 tons of fungi, worth 5 billion RMB (Mu et al., 2011). However, not all of these revenues came to local Tibetans, since in those years Cordyceps collection was not regulated, and many collectors came from outside TAR to profit from harvesting a fungus whose price had rocketed (ibid.).

Current Economic Development in Qinghai Province

China's production of *Cordyceps sinensis* reached 165.8 tons in 2021, of which Tibet accounted for 30.5% and Qinghai accounted for 61.6%, of which about 90% comes from Golog and Yushu Prefecture.[2]

Since China declared in 2021 that it had eliminated absolute poverty (Wenting et al., 2021), we refer to 2020 statistics to analyze the economic situation of each administrative region in Qinghai Province before the poverty eradication campaign ended. According to the 2020 statistics (Figure 11.2), Yushu Prefecture has the highest proportion of primary industry (over 60%), and the development of secondary and tertiary industries was relatively backward.

In contrast, the tertiary sector in Golog Prefecture, another major production area for Cordyceps, has become a major industry. In terms of GDP per capita and income per capita (Figure 11.3), Yushu was the region with the lowest per capita GDP in the province, and Golog Prefecture ranked second from the bottom.

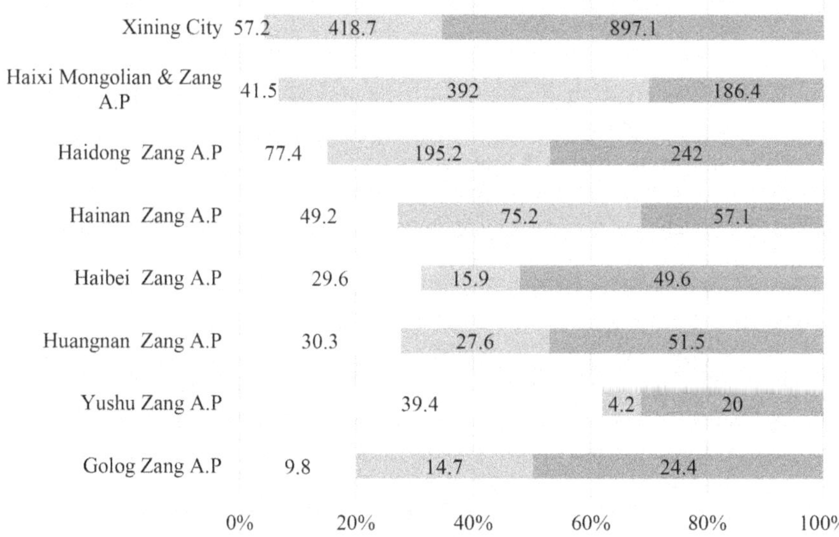

Figure 11.2 **GDP** distribution of the three industries for each region of Qinghai Province in 2020

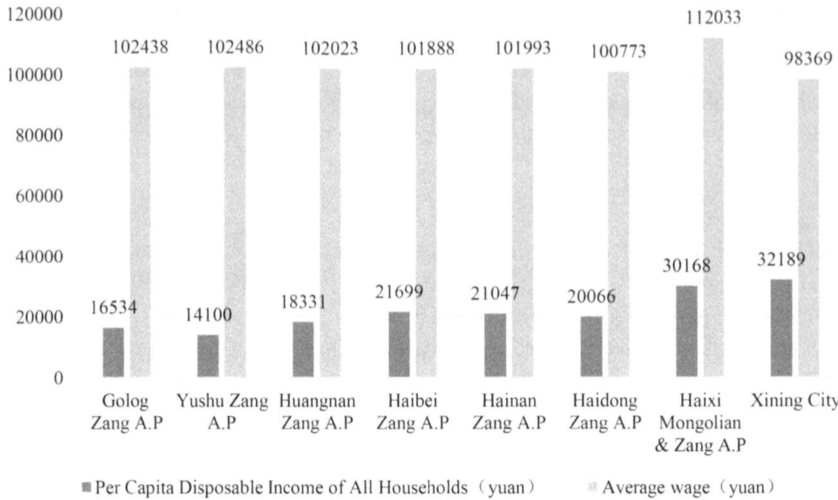

Figure 11.3 Per capita GDP of each region of Qinghai Province in 2020

Source: 2021 Qinghai Statistical Yearbook.

In addition, the introduction of the new ecological protection policy in the Sanjiangyuan region has also had a limiting impact on local agricultural production (Li et al., 2015; Tao, 2017). For example, the distribution of industrial output in Golog seemed to be similar to those economically developed areas such as Haibei Prefecture, but its GDP level was lower than Haibei's, probably indicating that Golog's industry might have been affected by the ecological protection project. This shows that pastoralists in the four counties are in a vulnerable state: on the one hand, they have lost some of their farming activities, and local industrial development is incomplete; on the other hand, reliance on collecting *Cordyceps sinensis*, which serves as the substitute income, is a risky and short-term strategy, as climate change and overharvesting can reduce the natural supply of the fungus.

Harvesting Cordyceps is not a direct guarantee of a comfortable life. According to the National Bureau of Statistics of China, the per capita consumption of Qinghai Province residents in 2022 was relatively good, and the Engel coefficient is lower than the national average. However, Qinghai also faces serious housing quality deficiency and lower living standards. Statistics from 2016 showed that 1% of villages in Qinghai Province did not have electricity, nearly 90,000 farmers had housing that was not safe, and 210,000 farmers did not have access to safe drinking water (Qinghai Statistical Information Network, 2018). Yushu and Golog Prefectures are the main production areas of *Cordyceps sinensis*, but the living conditions of the residents here were even less optimistic. The consumption level of rural residents in Golog and Yushu Prefectures was not high—far below the average level for Qinghai Province, and Engel's coefficient in Golog exceeded 50% in 2016, in fact reaching only the subsistence level.

Structure of Farmers' Income in Cordyceps-Producing Areas

We will further discuss the marginalization tendency in Qinghai-Tibetan agriculture, particularly in animal husbandry, which has been affected by the national ecological immigration protection policy, where a large number of farmers and herdsmen have been relocated. Against this background, harvesting *Cordyceps sinensis* has become the key income-producing industry in the region, which includes a million-strong army of local collectors of raw cordyceps and laborers who process Cordyceps in the distribution centers. With the rising prices, boosted by international and national demand and the rarity of the resource, the local *Cordyceps sinensis* economic industrial chain has provided necessary income to local residents, and helped to reduce the burden on local governments to subsidize highland agriculturalists (DuoJie, 2014; Pandey, 2022; Liu, 2022).

The income of farmers and herdsmen in 'the contiguous areas' of Tibet was still far lower than in other provinces and cities. Figure 11.4 shows the per capita net income of farmers and herdsmen in the main production areas of *Cordyceps sinensis* in 2016. It shows that the income level of the production areas was behind the average level of other administrative units. In particular, corresponding indicators in Yushu Prefecture and Golog Prefecture in Qinghai Province were lower than the national average income level, by more than 50%.

Family business net income has always dominated the per capita income of farmers and herdsmen, as it serves as their main source of income. An important contribution still comes from animal husbandry. In recent years, state subsidies have contributed to the income of farmers and herdsmen, and in Yushu Prefecture the proportion is as high as 30%.

Gannan Tibetan Autonomous Prefecture in southern Gansu also produces a certain amount of *Cordyceps sinensis*. A survey comprising 50 respondents was conducted to analyze sources of income and living conditions of farmers collecting

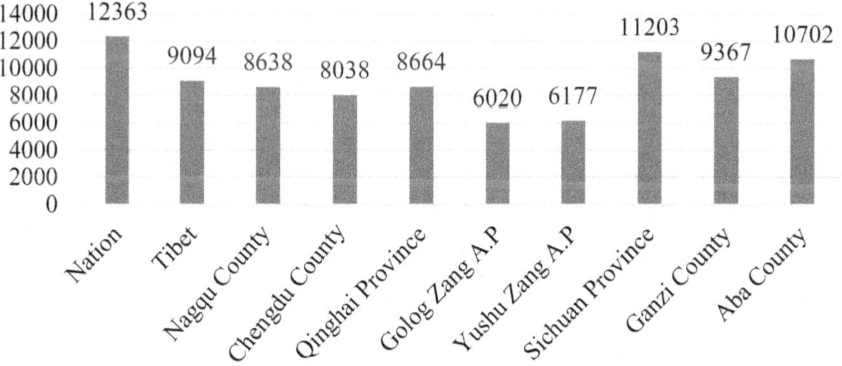

Figure 11.4 Per capita net income of farmers and herdsmen in the main production areas of *Cordyceps sinensis* in 2016

Source: Statistical Yearbooks 2017 for China, Tibet, and Sichuan, and Qinghai Statistical Yearbook 2010.[3]

Cordyceps sinensis in Sunan County, Gansu Province. Based on the answers received, the following classification of rural households was suggested: traditional agricultural households (relying only on farming activities), mixed agricultural households and non-farming households. The mixed households often included family members working in trade. To distinguish the different land rights, we divided 'pure agricultural households' into 'pure agricultural household I' (self-owned pasture use rights, referred to as property rights) and 'pure agricultural households II' (no property rights). The statistical results show that among the surveyed households, there were 29 pure agricultural households, of which group I and group II account for 46.34% and 24.39% of the total number of surveyed households, respectively. There were 13 households collecting *Cordyceps sinensis*, accounting for 31.71% of the total number of the surveyed households. See Table 11.1 for further details.

The classification of farmer households is mainly based on the proportion of agricultural income and non-agricultural income, in the economic income structure. According to the concept of agricultural income, crop income, animal husbandry income, and income from *Cordyceps sinensis* collection are classified as agricultural income; wage income, part-time work or business revenues are classified as non-agricultural income.

Among the three types: pure agricultural households, mixed households and non-farm households, a pure type of agricultural household is predominant. Through interviews with the village chiefs of agricultural and herders villages in Sunan County, the difference between the two was identified: herders of the animal husbandry villages let their animals graze in their own pastures. The number of migrant workers in the village is small; most of them are raising cattle. The agricultural villages are different, as the pastures are used by professional herders. The villagers can hand over their livestock to the professionals for marking and stocking. In order to maintain their livelihood, farmers work in the local area during the quiet season, relying on seasonal work, such as construction, mowing, and collecting *Cordyceps sinensis* to increase their income. According to this fieldwork, compared to the herdsmen's villages, the income of agricultural villages is relatively low.

Table 11.1 Basic information from the questionnaire conducted in Sunan County, Gansu Province

Statistics category	*Specific options*		*Number*	*Frequency*
Farmer type	Pure agricultural household	I (Property rights)	19	46.34%
		II (No property rights)	10	24.39%
	Agricultural mixed household		3	7.32%
	Non-farm household		9	21.95%
	Total		41	100.00%
Collect *Cordyceps sinensis*?	YES		13	31.71%
	NO		28	68.29%
	Total		41	100.00%

Through the data cross-analysis, a correlation between type of farmer and Cordyceps-collecting activity has been identified, and the economic income structure is the key factor determining the type of farmers who are excavators (Table 11.2).[4]

In pure agricultural households type I, pure agricultural households type II, mixed households and non-farming households, there were 52.63%, 20.00%, 0.00%, and 11.11% of farmers collecting *Cordyceps sinensis*, indicating that there is a positive correlation between the type of household and collection of *Cordyceps sinensis*. Among them, pure agricultural households I and pure agricultural households II comprised 12 households and were accounting for 92.31% of the total number of Cordyceps-producing households. The main reasons for this are:

(1) Pure agricultural households mainly rely on traditional agriculture and livestock income to maintain their livelihood. In addition to the labor of traditional agriculture, pure agricultural households will choose to obtain additional income during the quiet season.
(2) The high profits of *Cordyceps sinensis* stimulate the continuous harvesting by farmers.
(3) For non-farming and mixed households, most of their own economic sources are stable and the income is sufficient. If they choose to dig *Cordyceps sinensis*, the opportunity cost will be high.

These results might be compared with the study by Li Fen et al. that investigated *Cordyceps sinensis* collection activities in the Sanjiangyuan area.

Sanjiangyuan in Yushu County, Qinghai Province, has become part of the Chinese government's ambitions to create a large network of nature parks across the country by 2030, with the main aim of protecting endangered species and restoring degraded highland pastures. Conservation efforts were assessed by scientists (Obermann, 2020), and as a result, 10,140 households comprising of 50,000 herdsmen had to collectively relocate (Shi & Yang, 2023). The implementation project for the conservation park had two stages and was based on

Table 11.2 Sunan, Gansu: Analysis of types of farmers and *Cordyceps sinensis* digging activity

Farmer type		*Dig Cordyceps sinensis or NOT*				*Total*
		YES		*NO*		
Pure agricultural household	I (Property rights)	10	**52.63%**	9	47.37%	19
	II (No property rights)	2	**20.00%**	8	80.00%	10
Mixed household		0	0.00%	3	100.00%	3
Non-farming household		1	11.11%	8	88.89%	9
Total		13	/	28	/	41
Chi-Square Tests χ^2=7.631 df=3 Asymp.Sig. (2-sided) =0.054						

Note: 'χ^2' indicates the chi-square test value, 'df' indicates the degree of freedom, and 'Asymp.Sig. (2-sided)' indicates the significant probability of the two-tailed test.

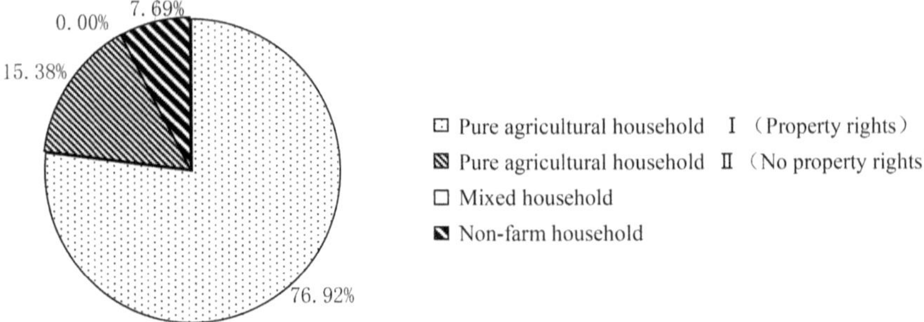

Figure 11.5 Sunan, Gansu: pie chart displaying the percentage of each household type engaged in the collection of *Cordyceps sinensis*

two policies—ecological compensation and ecological relocation—which were not clearly defined, resulting in the high dependency of relocants on governmental aid, and their continuous income instability (Ma et al., 2020). Shi and Yang (2023) reported that more than half of the ecological immigrants they surveyed said that they were able to adapt, but 40% of the immigrants experienced maladjustment. Relocants reported that they were able to continue their usual social, cultural and religious activities, but the key problem was economic well-being and stability due to their poor labor skills and few opportunities for employment in the new place.

As shown in Table 11.3, the collection of *Cordyceps sinensis* in Sanjiangyuan District had a higher contribution to the income of farmers and herdsmen. Among the farmers and herdsmen collecting *Cordyceps sinensis*, such income contributed 83.51% to the family business income. The income from collecting *Cordyceps*

Table 11.3 Contribution of *Cordyceps sinensis* collection to per capita household income of Sanjiangyuan farmers and herdsmen

	Total households researched			*Households with* Cordyceps sinensis *income*		
	Total (N=37)	*Herder (N=199)*	*Ecological immigrant household (N=174)*	*Total (N=278)*	*Herder (N=158)*	*Ecological immigrant household (N=120)*
Family business income (Yuan)	2761	3749	1438	3578	4808	1677
Per capita *Cordyceps sinensis* income (Yuan)	1565	2095	958	2988	2638	1389
Cordyceps sinensis income contributes to family business income (%)	56.68	55.88	66.62	83.51	54.87	82.83

sinensis in ecological migrant households contributed 27.96%, which was higher than among those who were practicing herders. This shows that the digging of *Cordyceps sinensis* has become the main source of income for families due to lack of employment opportunities after the migration.

A similar conclusion can be made for Yushu. At the end of 2021, the Yushu permanent population reached 418,400 people, including 346,600 rural residents; per capita disposable income derived from *Cordyceps sinensis* collection has accounted for up to 54.6% of total household income. By the end of June 2022, financial institutions provided a total of 928 million yuan for various types of *Cordyceps sinensis* loans, benefiting a total of 16,049 Cordyceps enterprises[5] of dealers and farmers (households), and the *Cordyceps sinensis* industry had become one of the important pillars of economic development in the state.

At present, the *Cordyceps sinensis* collection management accumulated funds and national help funds have been effectively combined to organize rural cooperatives to manage pastures and livestock, and the residents of the whole 'Cordyceps village' are entitled to receive cooperative shares through the dividend system, and further standardization and upgrading of Cordyceps collection and processing in some villages and townships in Yushu County in 2021 meant per capita dividends reached the 10,000 yuan target.[6]

Limiting Factors to Farmers' Income Increase

As discussed, although the price of *Cordyceps sinensis* is high and the yield is relatively large, the main production areas of *Cordyceps sinensis* have the deepest poverty and the widest inequality in income distribution. This can be explained by the influx of Cordyceps collectors coming from outside the region, as mentioned by interviewed farmers.

According to incomplete statistics, in 2006, there were more than 120,000 people—referred to as 'the digging army'—entering *Cordyceps sinensis*-producing pastures in Sanjiang. This 'army' was 6–7 times larger than the population of local farmers and herdsmen. The high value of the resource indicated that there might still be income growth potential for local farmers and herdsmen, especially in resource-rich and less populated areas (Galden, 2013).

With the recent resource governance system of *Cordyceps sinensis*, most highland villages with Cordyceps-producing pastures allow merchants to rent them(Fan, 2016). During the harvesting season, in the absence of effective trust, the contractor (merchant) often avoids hiring locals to collect *Cordyceps sinensis*. As a result, the income of farmers and herdsmen who do not have property rights over Cordyceps pastures has been rather mean. In most areas of Yushu, the collection permit system is still being implemented, and contracting pastures to outsiders is not allowed. Outsiders who enter the pastures and intend to collect *Cordyceps sinensis* must pay a certain fee to the village collective, while villagers with property rights can choose whether to collect it themselves or not, and the income of the village collective is distributed in a way that benefits those with property rights.

According to our surveys, the net income of the villagers who own the property rights over pastures that produce Cordyceps can reach more than 200,000 in one year under the collection certificate system. Often, as the main production areas of *Cordyceps sinensis* are mainly Tibetan people's settlements, language is a major barrier restricting farmers and herdsmen's income. The lack of Chinese Mandarin proficiency prevents effective communication and direct engagement in the sales and marketing of *Cordyceps sinensis*, so their benefits are minimal.

Such findings can be used in the discussion of how to lift poor Tibetan farmers out of poverty. In compliance with Sen's argument that loss of access and property rights makes people fall into poverty (Sen, 1983), the key recommendation is to clarify and protect the system of property rights and engage farmers in participating in the Cordyceps value chain management.

It would be useful to establish a unified management system for the collection of *Cordyceps sinensis*, to standardize procedures, and supervise relevant links in the Cordyceps production value chain. As we have shown, measures to regulate the *Cordyceps sinensis* market could improve the income and well-being of vulnerable groups.

Further Considerations that Would Alleviate Poverty and Promote Sustainable Development

Supply Chain Resilience

The *Cordyceps sinensis* supply and industrial chains are not resilient enough to deal with the impact of major crises. Under the impact of COVID-19, out-of-province Cordyceps dealers, especially wholesale customers from medium- to high-risk areas such as Beijing, Shanghai, Shenzhen, etc., had difficulties in entering Yushu Prefecture to buy Cordyceps, resulting in the local dealers in Yushu accumulating stocks, which affected their capital turnover. At the same time, the couriers who delivered Cordyceps directly to consumers could not enter certain medium- and high-risk areas, which affected the consumers' purchasing behavior. It resulted in dealers only being able to sell the *Cordyceps sinensis* to Lhasa, Yinchuan and other areas that were not so affected by the COVID-19 quarantine.

'*Cordyceps sinensis* dependence disease' is in fact an obstacle to diversifying rural industries. As many villagers in our 2023 interviews mentioned, 'Cordyceps can bring additional income' and 'the price of cordyceps determines the quality of life.' As a result of the high economic benefits of Cordyceps, farmers and herdsmen are seriously dependent on the *Cordyceps sinensis* economy. Some farmers and herdsmen regard Cordyceps collection as a sustainable job and overly rely on this fungus. At the same time, the wild *Cordyceps sinensis* economy is mostly dependent on natural resources, which does not effectively stimulate economic development, and there exists the phenomenon that 'some farmers and herdsmen are not willing to do anything or know how to do anything except digging cordyceps,' as mentioned by an authority representative in an interview in 2023.

The cascading role of the Cordyceps industry employment has been weakened. After years of development, wild *Cordyceps sinensis* harvesting and processing has gradually formed a stable industrial chain, driving the development of related industries and creating a large number of jobs in the Tibetan area. However, under the impact of COVID-19 as well as the commercialization of artificial *Cordyceps sinensis*, the decline in the natural production of wild *Cordyceps sinensis* and other multiple factors, the dependence on wild *Cordyceps sinensis* as the main source of income for farmers, herdsmen, merchants and downstream groups can create challenges for the stability of household income and rural residents' well-being in the future.

The distinction between wild *Cordyceps sinensis* and artificially cultivated *Cordyceps sinensis* is becoming increasingly blurred. According to Xining City *Cordyceps sinensis* market dealers, in 2023 in Golog, Ganzi and other *Cordyceps sinensis*-producing areas had already encountered the newcomer—artificial *Cordyceps sinensis*. A number of Traditional Chinese Medicine universities, research institutes and pharmaceutical companies have recently artificially cultivated *Cordyceps sinensis*. Artificial and wild are rather similar and can easily be mistaken, which can lead to counterfeiting, but the price is different (less than 10 yuan to more than 40 yuan per 0.4 gram).

In addition, rural households in Tibetan areas, after receiving access to the consumer goods market, developed excessive purchasing behavior that created problems with cash flow in farming families, and such a trend should be acknowledged in the poverty eradication policy.

Improvement of Farmers' and Herdsmen's Health

Collecting *Cordyceps sinensis* is a delicate process; it often breaks during its removal from the soil, and the spores are often detached. Such damaged fungi are often difficult to sell but are sold cheap. Thus, some herders choose to pick their own Cordyceps to eat, rather than buying it, to cook as a soup or grind into a powder to mix with powdered ginseng and other Chinese herbs to offer as medicine to the elderly or those with chronic conditions. Arguably, access to cheaper Cordyceps contributes to farmers' and herdmen's health and potentially reduces poverty-associated illnesses.

Education as Poverty Alleviation

As discussed earlier, Tibetans often suffered from lower proficiency in Chinese and lack of professional skills. This educational gap was the result of a deficiency in schooling. It was common practice to send Tibetan children to boarding schools in other provinces, such as Shandong. Advances in the Tibetan education system were often based on voluntary initiatives of Chinese and Tibetan teachers. Under the poverty eradication campaigns, the Ministry of Education of China started to train Tibetan school teachers to introduce cultural integration and advanced educational concepts (Ministry of Education, 2017). With their increased income, Tibetan farmers started to donate to improve pre-school education of local children. For example, in a highland village in Gonghe County, Tibetan Autonomous Prefecture of Qinghai Province, a local farmer with private property rights over pastures, lent his land to others so they could build a kindergarten on it, in 2012. As the project proved to be successful with local children, and the local authority received additional funding, the pre-school fees for local children have been waived (Infinitus China, 2020).

The booming industry of Cordyceps has attracted high-end Chinese universities to conduct field research and engage local actors in research and product development. Following the establishment of the 'Qinghai-Tibet Plateau Characteristic Resources Science Workstation of Sun Yat-sen University, CUHK and Moyu County, Hotan Prefecture, Xinjiang Province jointly established the 'Sun Yat-sen University Western Desert Transformation and Characteristic Resources Science Field Station' (Mei & Liu, 2008).

Digitalization of Cordyceps Supply Chain to Protect Farmers' Income

Currently, lack of business intuition and efficient promotion channels combined with low market awareness are barriers for Tibetan farmers in expanding their market. But there is a new tool to help farmers. The digital economy is characterized by wide dissemination and high penetration of information, and through short videos, live broadcasts, online searches, and other e-commerce modes, more comprehensive linkage of supply and demand in the *Cordyceps sinensis* market can be achieved.

By watching live broadcasts from production sites, consumers can appreciate the hard labor that goes into harvesting *Cordyceps sinensis* and also develop more confidence in the product they are buying. The digital network is low-cost for Cordyceps farmers to directly set up online platforms, and this way they avoid the problem of having to pay intermediary agents (merchants), who benefit the most from Cordyceps sales. The problem of monopoly can also be avoided.

The results of the digitalization of the market have been outstanding. The e-commerce transaction volume and online retail sales of agricultural products in Naidong District increased from 0 yuan in 2018 to 18.12 million yuan, 1.86 million yuan, and 12.0721 million yuan in 2022, respectively (Jintai Information, 2023).

Yuan Shufei, aka 'Cordyceps brother,' a general manager of Tibet Bright Onion, has shared his e-commerce entrepreneurship experience in Tibet:

> Everyone thinks cordyceps is expensive. However, we work directly with herders to de-intermediate, from the production area to the 'last mile,' to solve the problem of high yields but low returns for producers, high consumer prices and low value for consumers. This is also an important reason why Cordyceps can come out [to the public] . . .

He has also commented on how Tibetan traditional culture can be used for branding Cordyceps, to create a comparative advantage for Tibetan producers and bring about more just distribution of benefits:

Naidong is the birthplace of the Tibetan people and Tibetan culture. . . . Through platforms such as Douyin, Xiaohongshu, and Kuaishou, the customs and geography of 'Tibetan Source Naidong' have been widely exposed. The majestic Yala Shampo Snow Mountain, the rushing Yalong River, and the unique charm of Tashi Shepa Tibetan Opera . . .This arouses a lot of interest in this highland city. The 'original content + big data' closely connects suppliers of agricultural goods and potential consumers together, so they are constantly provoked by the desire to buy . . .

Now we have built a relatively complete online store matrix. With the help of public domain platforms such as Taobao, JD.com, and Xiaohongshu, as well as private domain platforms such as the 832 platform and State Grid, we can vigorously expand global e-commerce. Through e-commerce, live broadcast and other forms, 'Zangyuan Naidong' is more vividly displayed in front of the consumers. At the same time, we have also set up a direct store for Tibetan specialties, a special product exhibition hall, and a store-in-store for special products offline to further strengthen the brand experience . . .

(Jintai Information, 2023)

Notes

1 Data Source: Statistical Yearbooks 2023 for Tibet, Qinghai, and Sichuan.

2 Data Source: Official website of Qinghai Provincial Government, website: https://www.qinghai.gov.cn/dmqh/system/2021/12/11/010398983.shtml

3 National Bureau of Statistics (2017). Statistical bulletin on national economic and social development. https://www.stats.gov.cn/;

China Economic and Social Development Statistics Database (2017). Statistical analysis. http://tongji.cnki.net/kns55/Dig/dig.aspx;

Tibet Bureau of Statistics (2017). Qinghai Statistical Yearbook 2011–2017. China Statistical Press: Beijing, China;

Qinghai Bureau of Statistics (2017). Qinghai Statistical Yearbook 2011–2017. China Statistical Press: Beijing, China;

Sichuan Bureau of Statistics (2017). Qinghai Statistical Yearbook 2011–2017. China Statistical Press: Beijing, China.

4 This can be represented as χ^2=7.631, and the probability of significance is 0.054.

5 Data Source: Yushu news, website: https://www.yushunews.com/system/2022/08/01/013611919.shtml

6 Data Source: Official website of Qinghai Provincial Government, website: https://www.qinghai.gov.cn/zwgk/system/2024/02/02/030036549.shtml; China News Network, website: https://baijiahao.baidu.com/s?id=1729800201609478711&wfr=spider&for=pc

References

Bauer, K. (2005). Development and the enclosure movement in pastoral Tibet since the 1980s. Special Issue: Pastoralists in Post-socialist Asia. *Nomadic Peoples*, *9*(1/2), 53–81.

Bauer, K. (2015). New homes, new lives—the social and economic effects of resettlement of Tibetan nomads (Yushu, Prefecture, Qinghai Province, PRC). *Nomadic Peoples*, *19*(2), 209–220.

Chapela, L.R. (1992). Economic institutions of Buddhist Tibet. *The Tibet Journal*, *17*(3), 2–40.

DuoJie Zhaxi (2014). The educational game dilemma and strategy analysis behind the Western Cordyceps economy. *Journal of Anhui Agricultural Sciences*, 16, 5268–5269.

Fan, C-F. (2016). Economic form and cultural change of *Cordyceps sinensis* in Qinghai-Tibet area. *Folklore Studies*, 1, 118–128.

FAO (1999). *Towards a harmonised definition for non-wood forest products*. Unasylva.

Galden, T. (2013). *Study on ecological immigration in Sanjiangyuan District*. PhD dissertation, Shanxi Normal University.

Galden, T., & Li, Z. (2021). Tibetan ecological protection, resource development and farmers' income: Taking caterpillar fungus as example. *Tibetan Studies*, 5, 114–120.

Gama, D. (2020, December 18). The snowy plateau witnesses the miracle of anti-poverty—a record of Tibet's historic eradication of absolute poverty. *Guangming Daily*.

Goldstein, M.C., Childs, G., & Wangdui, P. (2008) Going for income in Village Tibet: A longitudinal analysis of change and adaptation, 1997–2007. *Asian Survey*, *48*(3), 514–534.

Ho, P. (2003). Mao's war against Nature? The environmental impact of the Grain-First campaign in China. *The China Journal*, 50, 37–59.

Infinitus China (2020). *Kindergarten on the Plateau* (高原上的幼儿园). https://www.infinitus.com.cn/c/2020-10-23/170642.shtml

Jeong, J. (2015). Ethnic minorities in China's Western development plan. *Journal of International and Area Studies*, *22*(1), 1–18.

Jintai Information (2023, September 13). 小虫草大蜕变 '数商兴农'为西藏乡村振兴闯出'金路子'(Cordyceps transforms into a 'golden way for Tibet's rural revitalization'). https://baijiahao.baidu.com/s?id=1776904751006094275&wfr=spider&for=pc

Kreutzmann, H. (2011). Pastoral practices on the move—recent transformations in mountain pastoralism on the Tibetan Plateau. In H. Kreutzmann, Y. Yong, & J. Richter (Eds.), *Pastoralism and rangeland management on the Tibetan Plateau in the context of climate and global change* (pp. 200–224). GIZ/GMZ: Bonn.

Li, H. (2015). Thoughts on targeted poverty alleviation in Tibet and Tibetan areas of four provinces. *The Theoretical Platform of Tibetan Development*, 6, 45–49.

Liu, Y., Zhou, S., & Chen, Y. (2022). How do local people value ecosystem service benefits received from conservation programs? Evidence from nature reserves on the Hengduan Mountains. *Global Ecology and Conservation*, 33, e01979.

Luorong, Z., & Dawa, T. (2006). Impact of *Cordyceps sinensis* resources on farmer and pastoralist income growth. *China Tibetology*, 2, 102–107.

Ma, T., Xu, K., Xing, Y., Shan, H., & Sang, W. (2020). Tendencies of residents in Sanjiangyuan National Park to the optimization of livelihoods and conservation of the natural reserves. *Sustainability*, 12, 5173. doi: 10.3390/su12125173

Mathou, T. (2005). Tibet and its Neighbors: Moving toward a new Chinese strategy in the Himalayan region. *Asian Survey*, *45*(4), 503–521.

Mei, Z., & Liu P. (2008, August 27). Sun Yat-sen University scientists grow *Cordyceps sinensis* in Tibet. *Nanfang Daily*.

Ministry of Education of the People's Republic of China (2017). 为雪域高原孩子插上隐形的翅膀 世界屋脊上的'内地西藏班' (Plugging in invisible wings for children on the snowy plateau 'Inland Tibet Class' on the roof of the world). www.moe.gov.cn/jyb_xwfb/xw_zt/moe_357/jyzt_2017nztzl/2017_zt06/17zt06_mtbd/201708/t20170830_312751.htm

Mu, W., Tashi, K., & Zhuoga, B.D. (2011) Contribution of *Cordyceps sinensis* to Tibetan pastoralist income and problems in its sustainable use. In H. Kreutzmann, Y. Yong, & J. Richte (Eds.), *Pastoralism and rangeland management on the Tibetan Plateau in the context of climate and global change* (pp. 165–176). GIZ/GMZ: Bonn.

Obermann, K. (2020, August 27). China forges ahead with ambitious national park plan. *National Geographic*.

Pandey, K. (2022). A review of the medicinal value of Ophio*Cordyceps sinensis* (Yarshagumba). *Asian Journal of Pharmacognosy*, *6*(1), 6–9.

Pradhan, B.K, Sharma, G., & Subba, B. (2020). Distribution, harvesting, and trade of Yartsa Gunbu (Ophio*Cordyceps sinensis*) in the Sikkim Himalaya, India. *Mountain Research and Development*, *40*(2), R41.

Qinghai Statistical Information Network (2018). The main data bulletin of the third national agricultural census in Qinghai Province (No. 4). https://www.qhtjj.gov.cn/tjData/survey-Bulletin/201802/t20180205_52781.html

Robin, F., & Hall, J. (2009). The 'Socialist New Villages' in the Tibetan Autonomous Region: Reshaping the rural landscape and controlling its inhabitants. *China Perspectives*, *3*(79), 56–64.

Sen, A. (1983). *Poverty and famines: An essay on entitlement and deprivation*. Oxford University Press: New York.

Shi, D., & Yang, Z. (2023). Sanjiangyuan eco-immigrants: Lifestyle, social adaptation, change of idea and urbanization process. *Frontiers in Sustainable Cities* 5, 1078153. doi: 10.3389/frsc.2023.1078153

Tao, W. (2017). *A comprehensive study on ecological security in centralized and poverty-stricken areas: Case studies in Sanjiangyuan poverty area*. Master's dissertation. Central China Normal University.

The Fourth National Survey on Chinese Materia Medical Resources (2017, October). *National Chinese herbal medicine price release*. Xining, Qinghai. https://www.zyzypc.com.cn/index.aspx?lanmuid=78&sublanmuid=685&id=12681

Wang, S. (2014). Challenges and opportunities for Tibetan farmers: Seeking non-farm income. *Asian Survey*, *54*(6), 1113–1135.

Wenting, X., Liu, C., & Shan, J. (2021, 25 February). China solemnly declares complete victory in eradicating absolute poverty. *Global Times*.

Wiley, T. (1986). Macro exchanges: Tibetan economics and the roles of politics and religions. *The Tibet Journal*, XI, 1.

Conclusions and Further Policy Recommendations

Jiping Sheng and Ksenia Gerasimova

This book is first a comprehensive summary of the *Cordyceps sinensis* industry and associated policies. We are pleased to have provided full details of morphology, habitat and medicinal quantities of the Ophiocordyceps, which we commonly called *Cordyceps sinensis* in the book. While it might serve as a small encyclopedia on this species, the key goal was to discuss the complexity of the supply chain and the impact of policies to make the industrial value chain more sustainable. In this final part, we address the recent challenges and constraints of the industry and discuss potential policy recommendations. In fact, studying a lucrative Cordyceps industry has served as a key to uncover complex issues in community development in the highland areas of Qinghai-Tibet.

The *Cordyceps sinensis* that we write about in this book is a wild, naturally grown fungus, which makes it a Non-Timber Forest Product. It is claimed that such products are an important source for invigorating developing economies and empowering rural communities (Ahenkan & Boon, 2010). While Cordyceps is not a basic necessity and not a staple crop for food subsistence, it has become the backbone of the rural economy in the Qinghai-Tibet region and even a wider area of the Himalaya. Rural farmers have partially or significantly given up on other rural activities, such as animal husbandry, which cannot compete in terms of revenues raised. From the analysis of the supply chain, we can see that *Cordyceps sinensis* has been successfully commercialized, yet its primary collectors, often Tibetan farmers, have small bargaining power and receive only a small premium on their harvest, with the main profit taken by traders and larger production enterprises. Having said that, we have also discovered that the middlemen in the Cordyceps trade are most at risk from market fluctuations, as they have to provide initial investment and coordinate with key processing enterprises.

A partial solution to increase bargaining power and share of the profit for farmers has been to unite into a cooperative and bypass the middlemen. With overarching digitalization, it has become a matter of having knowledge of technology and relevant digital platforms, which young Tibetans have been successfully demonstrating. These cooperatives do not create much competition for large-scale enterprises that supply processed Cordyceps for export or for large chains of pharmacies, but they might satisfy the demand of customers from the domestic market, who might appreciate the traceability of information and lower prices.

DOI: 10.4324/9781003429753-16

However, there is a significant factor that is endangering the prosperity of almost every actor in the supply chain of the Chinese Cordyceps industry—depletion of the natural pool of *Cordyceps sinensis* due to overharvesting and climate change. The issue of overharvesting was reported for the first time more than fifteen years ago. Due to high sensitivity to the conditions in its habitat, the natural productivity of *Cordyceps sinensis* has drastically fallen, reducing the fortune of its collectors, who can now bring home only a handful of these precious fungi. Under the current state of affairs, if there was no intervention, the species quickly would be on the brink of extinction, according to the recent modelling (Wei, 2021). The logical approach would be to strengthen the protection of this endangered species. China has already moved *Cordyceps sinensis* up the country's priority list of national protected species, and several nature reserves have been established in Cordyceps habitats. The regional authorities have regulated access to Cordyceps-producing pastures and controlled the harvesting practices and timing. Out of the four countries covered by the Himalayan Cordyceps habitat, China has the strictest control over harvesting. The Sanjiangyuan National Park protects the core area of the *Cordyceps sinensis* habitat, but local residents have lost their grazing pastures. It has been a necessary measure for biodiversity conservation; it was compensated but still requires monitoring and the implementation of further methods to help local farmers adapt to their new lives. Unfortunately, this is the reality of the combination of climate change and increased human activity in the Anthropocene.

Following the Porter Diamond model, our analysis shows that the international market for *Cordyceps sinensis* is on the brink of significant change. The most likely scenario is that *Cordyceps sinensis* will largely be produced artificially—i.e., in the laboratory. On one hand, this will remove the pressure on the fungus from overharvesting. On the other, the whole industry will be shaken, as several actors might become redundant at the procurement stage—that is, Tibetan rural collectors and Hui middlemen. At the moment, Chinese Cordyceps enterprises have a secure position in the domestic market and benefit from rising exports. However, as in the example of *Cordyceps militaris*, in the international market, large companies with headquarters in the USA have taken the lead, due to their technological advances in cultivating fungi in research facilities. Once commercialization of artificial *Cordyceps sinensis* is widely spread, new foreign entrants attracted by the ever-growing demand in Cordyceps and possibility to expand to North America and Europe might enter the international market. However, due to the established names of the Chinese firms and the reputation of their *Cordyceps sinensis*, they will likely retain their position in the Chinese domestic market and appeal to foreign consumers as the key providers of authentic natural Cordyceps with the best medicinal qualities.

More and more *Cordyceps sinensis* merchants and primary collectors are selling their *Cordyceps sinensis* on e-commerce platforms. Although the related risks have increased slightly, the development of e-commerce has intensified competition in the *Cordyceps sinensis* industry. Consumers can access multiple channels to learn about *Cordyceps sinensis* and compare prices, making the market increasingly transparent. The behavior of deceiving actors has decreased, and underhand dealings have become less common. The development of e-commerce has helped Tibetan farmers in remote production areas to participate in the sales of *Cordyceps sinensis* to the final consumers, so much so that they have started to diversify their sales by offering other agricultural and livestock products alongside caterpillar fungi. Local governments should use their credibility and promote the *Cordyceps sinensis* industry and support diversified sustainable agriculture. We have discussed in the book the origins of the 'Cordyceps rush' and how poor farmers have become dependent on this expensive fungus. While we assume that harvesting wild *Cordyceps sinensis* can be continued, it is important to work with farmers and herders to identify new potential jobs and support local initiatives to diversify their income-generating activities. For example, recently the Ministry of Traditional Chinese Medicine and the State Council Poverty Alleviation Office and five other departments jointly carried out the 'Chinese Medicine Industry Poverty Alleviation Plan' (2017–2020), which focused on reviewing policy tools for poverty alleviation. On the other hand, working with influential organizations or individuals from local communities is also a valid method to carry out a bottom-up needs appraisal.

We believe that enabling sustainable management of Cordyceps is still connected with the secured land use rights of herdsmen, who should be given another function—to serve as resource guardians and be offered ecosystem service payments in return. Considering climate change and potential market disruption, these payments will become an alternative to the soon-to-end boom of wild cordyceps harvesting. Such a policy has already been applied to grassland management in the country. In 2011, the Chinese government implemented a payments for ecosystem services (PES) policy, and in the following five-years overgrazing rates declined by more than 50% (GMSC, 2016).

Previous development programs for the West (Qinghai-Tibet) provided financial inputs for rural development and farming. While there was some criticism of the Chinese ecological resettlement schemes, these initiatives have supported farmers and herders. Thus, in the event of possible future climatic deterioration, it is important to assess and learn lesson from the previous schemes and include more comprehensive community-led elements when designing support programs. Local communities should also utilize science to increase resilience and disaster prevention methods under climate change. For example, planting trees can be helpful for soil erosion and to help prevent avalanches. Most likely, China will continue to implement such policy measures as ecological resettlement compensation, direct food subsidies, pasture subsidies, and provide infrastructure for agriculture and animal husbandry. Farmers and herdsmen in *Cordyceps sinensis* production areas should be encouraged to actively participate in the development of the *Cordyceps sinensis* industrial chain and related rural industries.

In the past, Chinese science engaged in campaigns to increase agricultural yields and animal stock. Lessons from this should be reviewed, and local communities should be given the lead in cooperation with scientific institutions. The establishment of agricultural and livestock research and innovation bonuses, and special loans for science and technology development and promotion, including mobilizing enthusiastic researchers and local personnel, will accelerate the development and protection of *Cordyceps sinensis*. Coordinating village-level construction and industrial development planning should solve the deficit of major facilities—roads, water, electricity, greenhouses, soil remediation, and construction of pastoral settlements—for the improvement of living conditions of farmers and herdsmen.

The delay in rural infrastructure development can be turned into an opportunity to construct greener infrastructure and apply green technologies. Qinghai-Tibet is rich in solar, geothermal and wind energy, which can compensate a lack of coal, oil and gas. We did not touch on this in the book, but Qinghai-Tibet could come to represent a good example of green energy development. In fact, UNDP has already helped to construct the largest geothermal plant in TAR. This could be used as a basis to protect the area from extractive activities which pollute the environment and do not benefit local residents. Since agriculture will remain the key activity of the local residents, it is advisable to implement agricultural practices that focus more on green and organic standards. It could be considered as a possible partial substitution for collecting wild Cordyceps in the future. It has been acknowledged that ecological products are economically important for rural livelihoods throughout the global South: their average annual contribution is estimated at 28% of the total family income (Angelsen et al., 2014).

We have already discussed in several chapters that *Cordyceps sinensis* collectors are herders and farmers who are confined to the front end of raw product excavation; the product is resold through intermediate agents, at a higher price to consumers, and revenue is suppressed, with the premium mostly generated in the intermediate circulation links. Buying it at low cost and selling it at a high price as agents do prevents rural cooperatives and individual front-end growers from taking the initiative in the market. With their lack of intuition regarding efficient promotion channels and their low market awareness, farmers find it difficult to expand into the larger market.

The digital economy is characterized by wide dissemination and high penetration, and through short videos, live broadcasts and shopping platforms, there is more comprehensive linkage between the supply and demand side of the *Cordyceps sinensis* market. Distance learning in the form of live broadcasts with good demonstrations can educate consumers about the product production scene and spark their interest in buying fresh fungi; this could likewise work as a channel to popularize science and disseminate information about the urgency of environmental protection and sustainable harvesting.

In the past, lack of education and business skills meant Tibetans could not enjoy the full benefits of either the developmental programs or the rising market for *Cordyceps sinensis*. With funds accumulated from the Cordyceps trade, new schools were built in remote locations, and such initiatives should be supported in the future. The government needs to encourage online platform enterprises to support agricultural science and technology research and development, empower traditional agriculture, improve agricultural production efficiency, increase income from quantitative production, and help with the digital transformation of agriculture. Intervening in all aspects of breeding, processing, brand marketing, logistics, and traceability to build the *Cordyceps sinensis* brand has changed the status quo and realized a win-win situation for poor households, consumers, and project implementation enterprises.

Live broadcasting and short videos have improved education, employment skills and digital literacy in rural areas, narrowing the divide. Tencent has transformed its technological capabilities to create digital social enterprises in various fields, such as industrial poverty alleviation, health poverty alleviation, education poverty alleviation, financial poverty alleviation, etc.

There are no readymade solutions to prepare actors in the *Cordyceps sinensis* supply chain for possible future climatic risks and market disruptions, but grounded optimism, long term-planning and active engagement of local communities can become working tools to ensure a good future for the people of Qinghai-Tibet and protected natural environment. Below are some possible policy recommendations in five strategic areas.

1. Promote the Scientific Upgrading and Efficiency of the *Cordyceps sinensis* Industry

- Focus on cultivating the industry's leading enterprises and supply them with key core technologies. Strengthen control of the whole industrial chain, key links and industry standards.
- Increase investment and support for scientific and technological capacity building, encourage cooperation between industry, Chinese academia and marketing research, and explore the development model of 'enterprise-initiated innovation + in-depth participation of scientific research institutes and colleges and universities + upstream and downstream of the industry chain + government support + development cooperation.'

2. Support the Development of the *Cordyceps sinensis* Industry to Reduce Information Asymmetry and Educate Collectors, Traders and Consumers

- Increase publicity for online market platforms and promote transparency.
- Engage locals' interest in nature conservation values.
- Create educational facilities; for example, a model of a Cordyceps pasture could be used to explain to collectors how to harvest without damaging the soil and the fungi left for reproduction.

3. Utilize *Cordyceps sinensis* Financial, Infrastructural and Labor Resources to Develop New Industries

- Support locally initiated cooperatives; promote organic agriculture and animal husbandry requirements and the premiums that can benefit farmers and herdsmen.
- Invest in developing tourism infrastructure.

- Encourage local farmers to value their cultural heritage and assist them in learning and commodifying traditional crafts, such as painting Tengka, pile embroidery, clay sculpture and stone carving, and lead them in understanding the management and marketing of the culture industry.
- Advise farmers and herdsmen to invest in local catering, accommodation, transportation, guided tours and other basic industries to improve the tourism industry.
- The Yushu yak industry can be used as a pilot project to create a free trade zone. The function of numerous nature parks created to protect vulnerable species can be re-assessed, and certain areas in the Sanjiangyuan National Park could be used for the development of eco paths. Local firms should be given the lead in developing tailor-made tours for Chinese and international tourists.
- Review so-called 'green mountains' and 'gold mountains,' in order to promote the organic unification of rural ecological construction and industrial development.

4. Strengthen Technical Research and Continuously Track the Development of Artificial *Cordyceps sinensis*

- Organize a national research group made up of researchers from the Chinese Universities of TCM, leading biochemistry experts and representatives of Chinese pharmaceutical enterprises to monitor the latest developments in artificial cultivation of *Cordyceps sinensis* and potential applications.
- Standardize the trade of *Cordyceps sinensis*, strengthen the supervision and management of the *Cordyceps sinensis* market, open up complaint and report channels, strictly investigate counterfeiting and the sale of fake products, and prevent illegal and irregular business behavior.

5. Make Education a Central Point in Poverty Alleviation Policy

- Ensure continuous investment in education, particularly to increase the proportion of public education expenditure.
- Provide basic education to all low-income groups through the implementation of compulsory education.
- Monitor management of rural education funds and encourage civil initiatives to support families most at risk of dropping out.
- Improve basic teaching facilities in rural primary and secondary schools; increase the number of teachers and strictly control the teaching quality.
- Strengthen the training of school masters and teachers.
- Improve the enrollment rate of students in agricultural and pastoral areas.

References

Ahenkan, A., & Boon, E. (2010). Commercialization of non-timber forest products in Ghana: Processing, packaging and marketing. *Journal of Food, Agriculture and Environment*, 8, 962–969.

Angelsen, A., Jagger, P.A., Babigumira, R., Belcher, B., Hogarth, N.J., Borner, J., Smith-Hall, C., & Wunder, S. et al. (2014). Environmental income and rural livelihoods: A global-comparative analysis, *World Development*, 64, 12–28. http://dx.doi.org/10.1016/j.worlddev.2014.03.006

GMSC (2016). *National grassland monitoring report* 2015 [in Chinese]. China Grassland Supervision Center, Ministry of Agriculture, P. R. China, Beijing.

Wei, Y., Zhang, L. Wang, J., Wang, W., Niyati, N., Guo, Y., & Wang, X. (2021). Chinese caterpillar fungus (Ophio*Cordyceps sinensis*) in China: Current distribution, trading, and futures under climate change and overexploitation. *Science of the Total Environment, 755*(1).

Index

For Product Safety Concerns and Information please contact our EU representative GPSR@taylorandfrancis.com
Taylor & Francis Verlag GmbH, Kaufingerstraße 24, 80331 München, Germany

www.ingramcontent.com/pod-product-compliance
Lightning Source LLC
LaVergne TN
LVHW010901110826
845149LV00005B/1436

* 9 7 8 1 0 3 2 5 5 2 5 2 1 *